THE RELATIVE MERITS OF THREE METHODS OF SUBTRACTION

An Experimental Comparison of the Decomposition Method of Subtraction with the Equal Additions Method and the Austrian Method

By

JOHN THEODORE JOHNSON, Ph.D.

TEACHERS COLLEGE, COLUMBIA UNIVERSITY
CONTRIBUTIONS TO EDUCATION, NO. 738

*Published with the Approval of
Professor Clifford B. Upton, Sponsor*

BUREAU OF PUBLICATIONS
TEACHERS COLLEGE, COLUMBIA UNIVERSITY
NEW YORK CITY
1938

ACKNOWLEDGMENTS

THE author desires, first of all, to express his indebtedness to his sponsor, Professor Clifford B. Upton, without whose valuable criticism and continued guidance and encouragement this study could not have been completed.

To his advisers, Professors W. A. McCall, Wm. D. Reeve, and John R. Clark, the author is especially grateful for their kindly help and giving of their time in reading the manuscript.

The author owes especial gratitude to Dr. Vera Sanford for critical reading of the historical material, and to Professor Helen M. Walker for helpful criticism on the statistical tables.

Acknowledgment is made to the eight principals, Leo Frederick, Jay R. Houghteling, Edna R. Myer, Jennie N. Phelps, Varian M. Shea, Kathryn E. Steinmetz, George White, and Robert I. White in whose schools the differential tests in subtraction were given. No less appreciation is extended to the teachers in these schools who suffered the inconvenience of having five tests at three different sessions given to their pupils. In this connection special mention should be made of Katherine M. Doyle, departmental teacher at the Lloyd School, in whose classes the Austrian method was studied and tested by some two hundred pupils.

Especial gratitude is extended to Elvira D. Campbell of the English faculty of the Chicago Normal College and to Professor W. W. Hatfield, editor of *English Journal,* for reading the manuscript.

Acknowledgment is also made to Professor Emeritus David Eugene Smith of Teachers College, New York, Dr. L. C. Karpinski of the University of Michigan, John C. Stone of State Teachers College, New Jersey, Dr. Vera Sanford of the State Normal School of Oneonta, N. Y., Dr. H. R. Hamley of London, England, Dr. Wm. Lietzman of Göttingen, Germany, and Professor A. Chatelet of Lille, France, for permission to quote from their various publications.

Finally, the author wishes to express his deep gratitude to his wife, Elsie Parker Johnson, for her patience and long suffering during the preparation, progress, and consummation of this study.

J. T. J.

CONTENTS

I

INTRODUCTION

The Problem

The specific problem of this study was to determine whether there is any difference in efficiency between the methods of subtraction in whole numbers as used in this country.

To the lay observer the methods of subtraction offer no question or challenge. To the child in school, to the teacher, and to the school administrator, however, the matter has long been a source of confusion. There is conflict even in the terminology describing the different methods. The term "additive subtraction," for example, may designate any one of three quite different procedures. Similarly, the term "borrow" is used with several meanings.

The confusion in method may be traced to several causes, frequently operating in combination. Authors of textbooks present in the same book two and sometimes three methods, lest a single method should fail to fit the course of study in certain cities or states. Teachers, finding two or more methods in the books they are using, choose the one they were taught—or the one they prefer. Principals hesitate to require the use of a uniform method in their schools, not knowing which to recommend since no one method has been scientifically proved to be superior. The consequence of all this is that in the various school systems in the United States there are to be found divers practices with respect to the teaching of subtraction. Pupils taught in one school or in one grade by one method may, by transfer or promotion, be brought into contact with other methods and thus fall into irrational and careless procedures.

A decided step toward improvement of this situation would be the adoption of some one method of subtraction. Yet little can be done toward this end until we have more conclusive evidence as to the relative merits of the three methods most commonly used in this country. It is the aim of the writer to present such evidence.

1

Methods of Subtraction Used in This Experiment

Because of the confusion in both terminology and procedure, the task of defining methods of subtraction becomes extremely difficult. Some authors say that there are three methods of subtraction; some say four, five, six, seven, and even nine. There are others who list twelve different kinds, and one author describes and illustrates thirty kinds of whole number subtraction. It will serve no purpose here to go into all the distinctions between the various methods and terms. The reader will find them discussed in Chapters II and III. It will be necessary, however, to make quite clear what the issue in this experiment is, and in order to do that the main methods of subtraction must be defined.

From the earliest times four main methods of subtraction in whole numbers have been taught in various parts of the world: the decomposition method, the equal additions method, the Austrian method, and the complementary method. The first three of these methods are used in this country and form the basis of this dissertation. It is conservative to say that they cover more than 95 per cent of the subtraction in whole numbers as taught in the United States today, for no other method is given in the arithmetic texts of the last twenty years. The fourth method, the complementary method,[1] very popular in Germany in the fifteenth and sixteenth centuries, is not used in whole numbers to any appreciable extent in this country; hence it will not form a part of this experiment.

The Decomposition Method

The person using the decomposition method in subtracting the example below thinks as follows:

$$
\begin{array}{ll}
81 & \text{``7 from } 11 = 4 \\
\underline{47} & 4 \text{ from } 7 = 3.\text{''} \\
34 &
\end{array}
$$

The method is explained by decomposing one 10 taken from the 8 of the minuend into ones and adding these ten ones to the 1 of the

---

[1] The complementary method is illustrated as follows:

$$
\begin{array}{ll}
81 & \text{The complement of 7 is 3} \\
\underline{47} & \text{3 and 1 are 4} \\
34 & \text{5 from 8 are 3, or 4 from 7 are 3.}
\end{array}
$$

units column of the minuend, making it 11. This is usually called the borrowing method, and is very common in the United States. Perhaps everyone who has had schooling past the third grade knows what borrowing means.[2] (The word "borrow" is used incorrectly here, for "borrow" suggests paying back what was borrowed to the source from which it came.)

The Equal Additions Method

The procedure by the equal additions method applied to the example below is as follows:

$$\begin{array}{r} 81 \\ 47 \\ \hline 34 \end{array}$$ "7 from $11 = 4$
    5 from $8 = 3$."

In this method ten units are added to the 1 in the minuend making it 11 and one 10 is added to the 4 of the subtrahend making it 5; hence the name "equal additions." The explanation is based on the principle that if the same number is added to the minuend and the subtrahend, the difference between them is unchanged.[3] (Many teachers erroneously attempt to explain this method by borrowing in the minuend and paying back in the subtrahend. Hence it is called by some the "borrow-and-pay-back" method. This is, of

[2] There is an additive variety of the decomposition method which differs from the subtractive variety explained above not in the manipulation of the figures but in the language of the user. In this additive variety the language used with the above example is "7 and 4 = 11, 4 and 3 = 7," the 4 and the 3 being the numbers sought. This is used by some students who think of their subtraction as addition with one addend missing. That is to say, when they subtract in a subtraction combination such as $\begin{array}{r} 15 \\ -8 \end{array}$ they say "8 and 7 = 15."

This method of thinking the simple subtraction combinations is referred to by many authors as additive subtraction, and the writer prefers to call it by that name in distinction to subtractive subtraction, which, when used in the above example, would be "8 from 15 = 7." Either the additive or the subtractive language may be used with the 100 simple subtraction facts.

Of the students tested in this experiment, less than ½ of 1% used the additive variety of the decomposition method. These few cases were not included in the study.

[3] There is likewise an additive variety of the equal additions method used by those who think "and" instead of "from" in their simple subtraction combinations. In the above example its users think, "7 and 4 = 11, 5 and 3 = 8." This is identical with the Austrian method in performance but differs from it in its explanation.

course, inaccurate and should not be sanctioned as an explanation.)
This method is used extensively in England and France at the present time.

The Austrian Method

A person using the Austrian method with the example below
thinks,

$$
\begin{array}{l}
81 \\
47 \\
\hline
34
\end{array}
\qquad
\begin{array}{l}
\text{``7 and 4 are 11} \\
\text{5 and 3 are \ \ 8.''}
\end{array}
$$

The 7 and 4 are thought of as addends, making 11. The left figure
1 of the 11 is carried as in addition to the 4 of the subtrahend, making
it 5. The 5 and 3 are thought of as addends again, making 8.
This method is often called the additive method,[4] which it truly
is. The author prefers, however, to confine the use of the term
"additive method" to the simple subtraction combinations mentioned in the footnote on page 3.

The Austrian method is the only method that is explained by the
reverse of addition and uses the additive language throughout.[5]

The experiment reported in this study compares: (1) the decomposition method with the equal additions method; (2) the
decomposition method with the Austrian method; (3) the equal
additions method with the Austrian method; and (4) the decomposition method (as the method that decreases the minuend figure)
with the equal additions and the Austrian methods combined (as the
methods that increase the subtrahend figure).

The four comparisons stated above will show differences in both
errors and time. To give added information, a further comparison
between the decomposition method on the one hand and the equal
additions and Austrian methods combined on the other will show

---

[4] See Stone, J. C. *How We Subtract*, p. 68. Benj. H. Sanborn & Co., Chicago,
1926.

[5] For assurance that those who were selected as using the Austrian method
were not using the additive variety of the equal additions method, the writer
took the precaution to use pupils of one school only where he was sufficiently
familiar with the situation to know that that method was the only one taught in
the third and fourth grades. All pupils from all other schools using this same
phraseology were eliminated. Fortunately there were only a very few of these.
This left as users of the equal additions method only those who used the subtractive variety as explained on page 3.

the difference in time only when the number of errors is constant.

The respective methods will be designated throughout the dissertation as D (decomposition), E (equal additions), and A (Austrian).

The description of the experiment and the details in the method of procedure, which is that of differential testing, are given in Chapter IV, page 41.

II

METHODS OF SUBTRACTION USED DURING THE PAST 800 YEARS

ILLUMINATING and interesting information on the confusion of methods and terms in subtraction may be gained from an examination of significant works on arithmetic in various periods.

THE COMPLEMENTARY METHOD

The complementary method will be considered first, as it was found in many of the older texts. David Eugene Smith says of it, "Bhāskara (c. 1150) used it—and no doubt it was even then an old one." [1]

In *The Crafte of Nombrynge*, an anonymous work based on the Carmen de algorismo of Alexander de Villa Dei (c. 1220), which was written in Latin verse, and one of the first arithmetics in English, we find the following statement concerning subtraction.

. . . behold an example $\frac{2122}{1134}$ take 4 from 2, it cannot be done, therefore borrow one of the next figure, that is 2, and set that over the first 2 and substitute it for ten. Where the second 2 stands write 1, for you have taken one from it. Then take the lower figure, that is 4, out of ten, and that leaves 6. Add to the 6 the figure of that 2 that stood under the caption of 1, that was borrowed and changed to ten, and that will be 8. Ignore that 6 and that 2 and set down 8 which let stand. When that is done, go to the next figure that is now but 1, but first was 2, whereof was borrowed 1. Then take from that the figure under it, that is 3. It cannot be done, therefore borrow of the next figure which is but one . . . [2]

---

[1] Smith, David Eugene. *History of Mathematics*, Vol. II, p. 98. Ginn and Company, Boston, 1925.

[2] " . . . lo an Ensampul $\frac{2122}{1134}$ take 4 out of 2, it wyl not be therefore borro one of the next figure, that is 2. and set that over the hed of the fyrst 2. & releue it for ten, and there (where) the secunde stondes write 1. for thou tokest on (one) out of hym. than take the nether figure, that is 4, out of ten. and then leues 6. cast to 6 the figure of that 2 that stode vnder the hedde of 1. that was

Note that the author borrowed 1 from the 2 tens not to change it to 10 ones and add it to the 2 ones, but to take 4 out of it.

The first arithmetic ever printed is known as the Treviso Arithmetic of 1478. It has no title and its author is unknown. It was printed in Italian in Treviso, a little town north of Venice. Smith has made a translation of part of it, which is published in his *Source Book in Mathematics*. The section describing the method of subtraction in whole numbers follows:

The number from which the other is subtracted is written above, and the number which is subtracted below, in convenient order, viz., units under units and tens under tens, and so on. If we then wish to subtract one number of any order from another we shall find that the number from which we are to subtract is equal to it, or greater, or less. If it is equal, as in the case of 8 and 8, the remainder is 0, which 0 we write underneath in the proper column. If the number from which we subtract is greater, then take away the number of units in the smaller number, writing the remainder below, as in the case of 3 from 9, where the remainder is 6. If, however, the number is less, since we cannot take a greater number from a less one, take the complement of the larger number with respect to 10, and to this add the other, but with this condition: that you add one to the next left-hand figure. And be very careful that whenever you take a larger number from a smaller, using the complement, you remember the condition above mentioned. Take now an example: Subtract 348 from 452, arranging the work thus:

$$452$$
$$348$$
$$\overline{}$$

Remainder 104

First we have to take a greater number from a less, and then an equal from an equal, and third, a less from a greater. We proceed as follows: We cannot take 8 from 2, but 2 is the complement of 8 with respect to 10, and this we add to the other 2 which is above the 8, thus: 2 and 2 make 4, which we write beneath the 8 for the remainder. There is, however, this condition, that to the figure following the 8 (viz., to 4), we add 1, making it 5. Then 5 from 5, which is an equal, leaves 0, which 0 we write beneath.

Then 3 from 4, which is a less from a greater, is 1, which 1 we write under the 3, so that the remainder is 104.[3]

---

borwed and rekened for 10, and that wylle be 8. do away that 6 & and that 2 & sette there 8, & lette the nether figure stonde stille. whanne thou hast do thus, go to the next figure that is now bot 1. but first yt was 2, & there-of was borred 1. than take out of that the figure vnder hym, that is 3. hit wel not be. ther-fore borowe of the next figure, the quych is bot 1 . . . " Steele, Robert. *The Earliest Arithmetics in English*, p. 12. H. Milford, London, 1922.

[3] Smith, David Eugene. *Source Book in Mathematics*, pp. 7, 8. McGraw-Hill Book Company, Inc., New York, 1929.

These two illustrations give two varieties of the complementary method, one diminishing the second minuend figure by 1, the other increasing the second subtrahend figure by 1. The latter variety is the one usually considered as the complementary method.

The complementary method was used by many writers in the fifteenth and sixteenth centuries. Among them were Adam Riese[4] in Germany, Gemma Frisius[5] in Belgium, and Humphrey Baker[6] in England. These three writers used the complementary method in whole numbers and in denominate numbers but not in mixed numbers.

L. L. Jackson[7] mentions the following as using the complementary method in whole number subtraction: Widman (1489), Pacioli (1494), Huswirt (1501), Tonstall (1522), Finaeus (1525), Tartaglia (1556), Trenchant (1566), Unicornus (1598).

The complementary method was used by Hodder in 1693 in mixed numbers and denominate numbers, but not in whole numbers. It was looked upon favorably by early American authors and was used in 1775 by Fisher[8] in mixed and denominate numbers and in 1788 by Nicholas Pike[9] in all phases of subtraction. Nathan Daboll[10] used it in mixed numbers and denominate numbers as late as 1837, but not in whole numbers. It was used only in denominate numbers by McCurdy[11] in 1850. This was as late as the complementary method was found by the writer in any American text. Fisher, Daboll, and McCurdy all used the equal additions method in whole numbers. Curiously enough, the word "borrow" continued to be used although these American authors did not borrow as did the author in the earlier source quoted from *The Crafte of Nombrynge*. Even Pike used the word "borrow," yet at

---

[4] Riese, Adam. *Rechenung nach der lenge auff den Linihen und Feder*. Leipzig, 1522 (1550 ed.).

[5] Frisius, Gemma. *Arithmeticae Practicae, Methodus Facilis*. Antwerp, 1540 (1581 ed.).

[6] Baker, Humphrey. *The Wellspring of Sciences*. London, 1568 (1591 ed.).

[7] Jackson, L. L. *The Educational Significance of Sixteenth Century Arithmetic*, p. 51. Contributions to Education, No. 8. Bureau of Publications, Teachers College, Columbia University, New York, 1906.

[8] Fisher, George. *The American Instructor*. John Bioren, Philadelphia, 1775.

[9] Pike, Nicholas. *A New and Complete System of Arithmetic*. Newburyport, Mass., 1788.

[10] Daboll, Nathan. *School Master's Assistant*. Ithaca, New York, 1797 (1837 ed.).

[11] McCurdy, D. *New American Order of Arithmetic*. Baltimore, 1850.

the same time he increased the subtrahend figure. He consistently used this method of increasing the subtrahend figure in all three phases of subtraction.

The following quotation will illustrate Pike's method in whole numbers:

> . . . but if the lower figure be greater than the upper, borrow ten and sub-tract the lower figure therefrom: to this difference, add the upper figure, which, being set down, you must add one to the ten's place of the lower line for that which you borrowed: and then proceed through the whole.[12]

Edmund Wingate, an English author, in 1760 used both varieties of the complementary method in mixed numbers. As an illustration of how this method is used in mixed numbers the following is quoted from him:

> Thus, if 20 7/8 be given to be subtracted from 35 3/5, the remainder will be found 14 29/40, viz. subtracting 7/8 from an integer, or 1, the remainder is 1/8, which added to 3/5, gives 29/40; then adding the integer borrowed to 20, it will be 21, which subtracted from 35, the remainder is 14, so the remainder or difference sought, is 14 29/40.[13]

In whole numbers and denominate numbers he employed the equal additions method but used the word "borrow."

As an illustration of the complementary method in denominate numbers an example is given from Humphrey Baker,[14] an English author of the sixteenth century.

li	*s*	*d*	
28	13	9	11 from 12 (because 12 pence is a shilling) is 1, 1 and 9 is 10,
15	17	11	17 becomes 18. 18 from 20 (because 20 shillings in a pound)
12	15	10	is 2, 2 and 13 is 15; write 15, 15 becomes 16. 16 from 28
			leaves 12.

There is a logical connection between the complementary method and the equal additions method which should be pointed out here. Note the slight difference between the two methods in the following example:

---

[12] Pike, Nicholas. *A New and Complete System of Arithmetic.* Newburyport, Mass., 1788.

[13] Wingate, Edmund. *Arithmetick, Containing a Plain and Familiar Method for Attaining the Knowledge and Practice of Common Arithmetick*, p. 139. Edited by James Dodson, N. P. London, 1629 (1760 ed.).

[14] Baker, Humphrey. *The Wellspring of Sciences*, pp. 13, 14. Thomas Purfoote. London, 1568 (1591 ed.).

42   The complementary procedure is: 7 from 10 is 3
27   3 and 2 are 5
——   3 from 4 is 1
15

The equal additions procedure is: 2 and 10 are 12
                                   7 from 12 is 5
                                   3 from 4 is 1

The difference is that the complementary method takes **7** from **10** first and then adds the **2**, while the equal additions method adds the **2** to the **10** first and then subtracts **7** from **12**.

Perhaps the reason the complementary method is now becoming extinct* is the fact that the equal additions method is more clearly seen and can be more easily presented. The writer has not found in the literature examined any logical explanation of the complementary method. It is, however, readily explained on the principle of equal additions, the amount added to minuend and subtrahend being the complement (with respect to 10) of the right-hand figure of the subtrahend instead of 10, as is the case in the usual equal additions method.

In the example at the left, in adding **3** (the complement of **7**)

42 becomes 45   to the **2** it becomes **5**. In adding 3 to the **7** it be-
27 becomes 30   comes 10, making the next figure 2, in the sub-
——   ——   trahend one more, or 3. See illustration for further comparison:

$$42 \quad 42 + 10 = 40 + 12$$
$$\underline{27} \quad \underline{27 + 10} = \underline{30} + \underline{7} \quad \text{(Equal additions)}$$
$$\qquad\qquad\qquad\qquad 10 \quad 5$$

$$42 \quad 42 + 3 = 45$$
$$\underline{27} \quad \underline{27 +} 3 = \underline{30} \quad \text{(Complementary)}$$
$$\qquad\qquad\qquad\qquad 15$$

THE DECOMPOSITION METHOD

The decomposition method in this country is commonly called the borrowing method. Smith speaks of it as follows:

* The technique of the complementary method is used in finding cologarithms, and it forms the basis of the subtraction in many calculating machines. Although now almost extinct in this country, this method may yet come into its own in connection with mixed number and denominate number subtraction.

The plan of simple borrowing is the one in which the computer says: "7 from 12, 5; 2 from 3 (instead of 3 from 4), 1." This method is also very old. It appears in the writings of Rabbi ben Ezra (c. 1140), the computer being advised to begin at the left and to look ahead to take care of the borrowing. This left-to-right feature is Oriental and was in use in India a century ago. It was the better plan when the sand table allowed for the easy erasure of figures, but it had few advocates in Europe.

$$\begin{array}{r} 42 \\ 27 \\ \hline 15 \end{array}$$

When the computation began at the right the borrowing plan was also advocated by such writers as Gernardus (13th century?), Sacrobosco (c. 1250), and Maximus Planudes (c. 1340). The writers of the early printed arithmetics were not unfavorable to it, although they in general preferred the borrowing and repaying method.[15]

Karpinski reports in *Isis* two twelfth century algorisms written in Latin.[16] The section dealing with subtraction clearly describes the method as that of decomposition.

Three other algorisms are reported in later issues of *Isis*. One of the thirteenth century[17] is in French verse as translated from the Latin. Another of the fourteenth century[18] is in Anglo-Norman verse and another of the fifteenth century[19] is in French verse. In all of these algorisms the subtraction is carried on by the method of decomposition.

The present writer found the decomposition method in a work of Levi ben Gerschom, dated 1321 and translated from the Hebrew into German by Dr. Gerson Lange.[20] Lange's translation is here given to show the use of the decomposition method in connection with sexagesimal fractions. Sexagesimal fractions, as the name implies, were fractions whose ratio between orders from right to left was 1 : 60 instead of 1 : 10 as it is in our decimal order or system. As two figures could occupy any order, each order or rank had to be separated from the adjacent ranks by spaces, as is noted in the example below. This ratio did not apply to the integral part

[15] Smith, David Eugene. *History of Mathematics*, Vol. II, pp. 100-101. Ginn and Company, Boston, 1925.

[16] Karpinski, L. C. "Two Twelfth Century Algorisms." *Isis*, Vol. III, pp. 396-413, 1921.

[17] Waters, E. G. R. "A Thirteenth Century Algorism in French Verse," with Introduction by L. C. Karpinski. *Isis*, Vol. XI, pp. 45-84, 1928.

[18] Staubach, Charles N. "An Anglo-Norman Algorism of the Fourteenth Century," with Introduction by L. C. Karpinski. *Isis*, Vol. XXIII, pp. 121-152, 1935.

[19] Waters, E. G. R. "A Fifteenth Century French Algorism from Liége." *Isis*, Vol. XII, pp. 194-236, 1929.

[20] Lange, Gerson. *Die Praxis des Rechners*. Frankfort, 1909.

of the number, which is that part to the left of the vertical bar, as shown in the example. These fractions were used by the Babylonians and the Greeks in scientific work and later by scientific writers of the Middle Ages. They were found in arithmetics as late as 1640.

Following is the example from Gerschom, with the translation of the subtraction process from Lange:

01079	45

| 31080 | 0 | 46 | 35 | 47 | 0 | 53 |
| 206 | 50 | 0 | 37 | | | |

| 30873 | 10 | 45 | 58 | 47 | 0 | 53 |

Then we subtract from the 35 in the upper row that which stands directly under it in the lower, that is 37, that we cannot subtract from 35, therefore we take one out of the heading which comes after 35, in this rank it will be 60, adding it to 35 makes it 95, we subtract therefrom 37, thus leaving 58 which we write in the row of the result under the heading Terzen; hence leaving only 45 left in the following order.[21]

L. L. Jackson [22] mentions the use of this method in the sixteenth century by Pacioli (1494), Köbel (1514), Tartaglia (1556), Buteo (1559), Champenois (1578), Raets (1580), and Ramus (1586). Pacioli and Tartaglia used also the complementary method, as previously related.

The origin of the word "borrow," no doubt, goes back to the abacus which was used for computation for several centuries. Information on this comes from Karpinski in *The History of Arithmetic*, wherein he says:

In adding two numbers on the abacus when you have ten stones in any one column you take them up and *carry one* stone over to the next column.

[21] "Beispiel:

"Wir wollen 206 Ganze, 50 Primen, 37 Terzen, von 31080 Ganzen, 46 Secunden, 35 Terzen, 47 Quarten, 53 Sexten abziehen.

01079	45

| 31080 | 0 | 46 | 35 | 47 | 0 | 53 |
| 206 | 50 | 0 | 37 | | | |

| 30873 | 10 | 45 | 58 | 47 | 0 | 53 |

"Dann Ziehen wir von den 35 in der oberen Reihe das ab, was in der untern gegenuber steht, das sind 37, das konnen wir nicht von 35 abziehen, deshalb *nehmen*[*] wir eines aus der Rubrik, die 35 kommt, es wird in dieser stufe 60, addiere sie zu 35, so sind es 95, ziehen wir davon 37 ab, so bleiben 58, wir schreiben sie in die Reihe des Resultats in die Rubrik der Terzen. Daher bleiben in der folgenden Rubrik nur 45 ubrig."

[*] The word "nehmen" is italicized by the present writer to call attention to the fact that this early author did not use the word borrow, as *nehmen* means to take.

[22] Jackson, L. L. *The Educational Significance of Sixteenth Century Arithmetic*. Contributions to Education, No. 8. Bureau of Publications, Teachers College, Columbia University, New York, 1906.

Or in subtraction you may *borrow one* stone from the next column to the left to make ten in the right-hand column.[23]

The Greeks and Romans used stones on the abacus, the Chinese and Russians used beads on wire frames, and later medieval Europe used loose counters thrown on lines drawn on a board or table. Counter reckoning was a common form of computation in medieval times.

Quoting again from Karpinski, we quote the following paragraphs:

The use of the abacus led to another type of computation, called *reckoning on lines.*

Little round markers or counters were made to use upon lines drawn upon a table called in German *ein Rechenbanck.*

A counter placed upon the first line represented a unit; placed upon the second line a ten; and so on. A counter placed in any space represents five counters in the line just below the space.[24]

Karpinski says also:

The "reckoning on lines" began in the thirteenth century and extended over all Europe. Long after the invention of printing, treatises on this subject continued to appear, usually together with the written arithmetic.[25]

Sanford makes the following statement on counter reckoning: (Italics are the writer's)

The influence of operations with the abacus on later computation with numerals makes it desirable to illustrate this work in some detail.

The loose-counter abacus may be easily reproduced. A piece of paper marked with four or five horizontal lines serves as a counting board. In medieval times this would have been a table with the lines checked or painted across it, or it might have been a piece of fabric with the lines woven or embroidered in the proper pattern. Coins may be used to represent the counters. In former days these would have been flat disks of copper, silver, or gold, according to the rank of the owner. Counters embossed with symbolic designs were used as souvenirs of important occasions and bags of gold counters were the proper New Year's gifts for princes. In fact, Louis XV regularly had a half-dozen gold plates made from the counters he received each year.

To add two numbers, as 282 and 369, the counters are placed as in the

[23] Karpinski, L. C. *The History of Arithmetic,* p. 26. Rand McNally and Co., Chicago, 1925.
[24] *Ibid.,* pp. 33, 34.
[25] *Ibid.,* p. 36.

accompanying figure, in which the numbers to be added and the sum are both represented although in actual work this would not be the case. Since

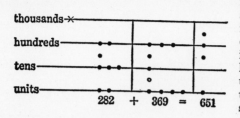

282 + 369 = 651

five counters on the units line are equivalent to one in the fives' space, five of the unit counters are removed and one is *carried* to the next space, leaving one counter on the units' line in the sum. The two counters now in the fives' space are equal to one counter on the tens' line. Accordingly, the two counters are removed and one counter is carried to the next line. The process is repeated until the sum remains in its simplest form. It is clear that the only skill required is to know how to lay out the counters and to be able to replace five counters on a line by one in the next space or to put one on the next line for every two that appear in a space. In subtraction, the subtrahend is literally *taken away* from the minuend and whenever any line or space in the minuend has an insufficient number of counters, others are *borrowed* from the space or line next above.[26]

Köbel's arithmetics gave detailed expositions of counter reckoning, a mechanical means of computation in which flat counters were laid on the lines and spaces of a counting board.[27]

Jackson says this relative to the origin of the word "borrow":

In those works which used the plan of supplying 10 from the next order of the minuend to make subtraction possible, one naturally seeks to find a trace of the modern vulgarism "to borrow" and recognizes it in the word "entlehen" used by Köbel. This is suggestive, because Köbel was primarily an abacist, and he would probably employ the same word in the algorism that was employed to describe the actual borrowing process in abacus reckoning.[28]

Thus it is quite certain that the abacus is responsible for the use of the word "borrow" in the literature of subtraction today.

Let us next trace the use of the decomposition method in the United States. It seems to have been introduced here by Warren Colburn in 1821,[29] gradually displacing the equal additions method previously prevalent. For more than forty years Colburn's and

[26] Sanford, Vera. *A Short History of Mathematics*, pp. 89, 90. Houghton Mifflin Company, Boston, 1930.

[27] *Ibid.*, p. 34.

[28] Jackson, L. L. *The Educational Significance of Sixteenth Century Arithmetic*, p. 53. Contributions to Education, No. 8. Bureau of Publications, Teachers College, Columbia University, New York, 1906.

[29] Colburn, Warren. *First Lessons in Intellectual Arithmetic.* Houghton Mifflin Company, Boston, 1821.

Emerson's [30] books were the only texts of any note that used it.[31] The word "borrow" was not used by either, but appeared several years later.

Colburn gives this explanation:

How many are 53 less 18?

$$53 = 5 \text{ tens} + 3 \text{ units}$$
$$18 = 1 \text{ ten } + 8 \text{ units}$$

We cannot take 8 units from 3 units; we therefore take one of the 5 tens and add it to the 3 units.

$$53 = 4 \text{ tens} + 13 \text{ units}$$
$$18 = 1 \text{ ten } + 8 \text{ units}$$

$$53 - 18 = 3 \text{ tens} + 5 \text{ units} = 35 \text{ Answer.}[32]$$

Note that Colburn did not use the word "borrow."

In the time of Colburn and Emerson object teaching was the order of the day, and teaching procedures were rationalized through objects. For example, in subtracting 27 from 42, four bundles of splints of ten each and two single ones were laid out. When 7 could not be taken from 2, one of the four bundles of ten would be opened up and the splints put with the two, making twelve; then 7 could be taken from 12, leaving 5, and since one of the four bundles was actually gone, the two tens had to be taken from the three bundles remaining, leaving one bundle of ten splints and five single splints, a total of fifteen.

$$\begin{array}{r} 42 \\ 27 \\ \hline 15 \end{array}$$

The change to this objective procedure seemed justifiable, and by 1875 many of the texts in the United States were using the decomposition method. There was no question whether or not the new method was more efficient than the old one. As long as it could be explained objectively it was satisfactory. The equal additions method had been given no logical explanation in most texts at this time other than that afforded by the term "borrow and repay," and it was losing ground.

In favor of the decomposition method was the fact that mixed number subtraction and compound denominate number subtraction could be easily explained and rationalized by this method.

[30] Emerson, Frederick. *North American Arithmetic*. Russell, Odiorne, and Metcalf, Boston, 1834.

[31] Stone, J. C. *How We Subtract*, p. 45. Benj. H. Sanborn & Co., Chicago, 1926.

[32] Colburn, Warren. *First Lessons in Intellectual Arithmetic*, p. 74. Houghton Mifflin Company, Boston, 1821.

Colburn's books consistently used it in the three places where such subtraction is employed. This uniformity of method in the three phases of subtraction was rarely accomplished in any of the older books.

THE EQUAL ADDITIONS METHOD

Concerning the equal additions method David Eugene Smith says:

The borrowing and repaying plan, in which the 1 that is borrowed is added to the next figure of the lower number, is one of the most rapid of the methods in use today and has for a long time been one of the most popular. It appears in Borghi's (1484) well known work, the first great commercial arithmetic to be printed. Borghi takes the annexed example and says, in substance:
6354   "8 from 14, 6; 8 from 15, 7; 10 from 13, 3; 3 from 6, 3." The plan
2978   was already old in Europe, however. Fibonacci (1202) used it, and
——   so did Maximus Planudes (*c.* 1340). These writers seem to have in-
3376   herited it from the Eastern Arabs, as did the Western Arab writer, al-Qalasâdi (*c.* 1475). . . . This method of borrowing and repaying was justly looked upon as one of the best plans by most of the 15th and 16th century writers, and we have none that is distinctly superior to it even at the present time.[33]

There is no evidence from Smith's account that these authors spoke of their methods as "borrowing and repaying." That may have been a name acquired later and applied to the earlier methods.

L. L. Jackson mentions the following sixteenth century authors as using the equal additions method: Piero Borghi (1484), Pacioli (1494), Tonstall (1522), Rudolff (1526), Cardan (1539), Noviomagus (1539), Tartaglia (1556), Van der Scheure (1600).[34]

In Tacquet's arithmetic of 1665, the example explained first by the decomposition method was also explained by the equal additions method. The translation follows:

8068 K   Given L, to be taken from K. 6 from 8 leaves 2. I cannot take
576 L   7 from 6, I therefore add 10 and I subtract 7 from 16: 9 remains.
——   Now because I have added 10, the following lower figure 5 is
7492 M   to be increased by unity, and thus becomes 6. Again I cannot take 6 from 0. I therefore take 6 from 10 and 4 remains. Because I have really added 10 again, I replace unity in the following place which is vacant, which taken from 8 leaves 7. The remainder sought is therefore M.

[33] Smith, David Eugene. *History of Mathematics*, Vol. II, pp. 99-100. Ginn and Company, Boston, 1925.

[34] Jackson, L. L. *The Educational Significance of Sixteenth Century Arithmetic*, p. 51. Contributions to Education, No. 8. Bureau of Publications, Teachers College, Columbia University, New York, 1906.

This method is more efficient and superior: it will be still more useful when several continuous zeros occur in the greater number.[35]

Note the judgment in favor of the equal additions method and also the fact that the word "borrow" does not occur. The translation of " 7 ex 6 subduci nequit. Adiicio ergo 10" is "I cannot take 7 from 6; I therefore add 10."

Then follows a geometrical explanation of the equal additions principle. It is reproduced here as one of the early explanations of this method giving the true name to the method.

E———A———L———B
I———C———F

If CF is to be taken from AB and equal quantities, EA and IC be added to AB and CF, the remainder, LB, will be the same as when CF was taken from AB.[36]

Had this explanation reached other countries earlier, it might have counteracted the misnomer of "borrow and repay." Unfortunately, the incorrect terminology has persisted.

In a posthumous edition of Edmund Wingate's *Arithmetick*,[37] the editions of which numbered over a hundred, is found the equal additions method in whole number and denominate number subtraction, but the phrase "borrowing 10 of the next rank" is used. It is strange that the word "borrow" should be used here since in the same edition the principle of equal additions is stated as follows and is given as the very reason for the procedure:

If to each of two numbers, the same or an equal number be added, the difference between the numbers resulting will be the same with the difference of the first two numbers.[38]

[35] "Modi II Exemplum

"Detur, L, auferendus ex K. 6 ex 8 relinquit 2. 7 ex 6 subduci nequit. Adiicio ergo 10, & 7 demo ex 16: restant 9. Iam quia adieci 10, sequens nota inferior 5 augenda est unitate, & sic 5 fit 6. Rursum 6 ex 0 auferri nequit. Aufero igitur 6 ex 10, & restant 4. Quia vero rursum adieci 10, repono sequenti loco, qui vacar, unitatem quae, ablata ex 8 relinquit 7. Residuum ergo quaesitum est M."

8068 K
 576 L
————
7492 M

"Haec methodus expeditior plerumque est praecedenti; quae tamen commodior erit, cum in maiore numero plures occurrent cyfrae continux."

[36] Tacquet, André. *Arithmeticae, theoria et praxis*, p. 112. I. Meursium, Antwerp, 1665.

[37] Wingate, Edmund. *Arithmetick, Containing a Plain and Familiar Method for Attaining the Knowledge and Practice of Common Arithmetick*, Edited by James Dodson, N. P. London, 1629 (1760 ed.). This is known as the nineteenth edition of Wingate.

[38] *Ibid.*, p. 24.

The notion that 10 could simply be added to both minuend and subtrahend did not seem to occur to the authors. Before they added the 10, they had to "borrow" it from somewhere.

The Oxford Dictionary gives the following as one definition of the word "borrow":

> Borrow . . . . c. Arith. in subtraction. When the number to be subtracted in one denomination is greater than that of the minuend, To transfer to the latter mentally the equivalent of a unit of the next higher denomination, compensating or "paying back" for this at the next step in the process.[39]

Evidently long usage of the term in this sense had received the sanction of the authorities in England during the nineteenth century.

In an arithmetic in which poetry abounded, "refined" by Nicholas Hunt, M. A., in London (1633), the following reference to borrowing occurs:

> Subtract the lesser from the great, noting the rest
> As ten to borrow you are ever prest,
> To pay what borrowed was thinke it no paine
> But honesty redounding to your gain.[40]

The common use of the word "borrow" in connection with the equal additions method is well exemplified by the following note from Daniel Adams, an American author:

> *Note.* The reason of *borrowing ten* will appear if we consider, that, when two numbers are equally increased by adding the same to both, their difference will be equal. Thus, the difference between 3 and 5 is 2; add the number 10 to each of these figures (3 and 5) they become 13 and 15, still the difference is 2. When we proceed as above directed, we add or suppose to be added, 10 to the *minuend* and we likewise add 1 to the next higher place of the *subtrahend,* which is just equal in value to 10 of the lower place.[41]

Thus the word "borrow" seemed to have acquired a new meaning when used in connection with subtraction. This is unfortunate for students and teachers using the word do not seem to be able to detach the common meaning from it. We are all subject to the tyranny of words.

The phrase "borrow and repay" very probably had its origin

---

[39] Murray, James A. H. *A New English Dictionary,* Vol. I, A and B, p. 1006. Oxford University Press, Oxford, England, 1888.

[40] De Morgan, Augustus. *Arithmetical Books from the Invention of Printing to the Present Time,* p. 40. Taylor and Walton, London, 1847.

[41] Adams, Daniel. *The Scholar's Arithmetic,* Revised Edition, p. 19. John Prentiss, Keene, N. H., 1815.

with teachers who felt it a duty to pay back what they borrowed. Stone says:

> The terms "borrowing" and "paying back" were largely devices invented by the teacher. I can find no standard author using these terms in the sense in which they were used by many teachers.[42]

In the writer's opinion the phrase "borrow and repay" should be carefully avoided. Not only has it caused confusion in thought and procedure, but it has also induced unfavorable emotional reaction on the part of some teachers. The following is quoted from Stone:

> Long after the decomposition method was introduced, many teachers clung to the older method. The author heard many amusing debates over the two methods thirty or forty years ago. Teachers would condemn the older method as being "immoral and unethical, for it taught the child to borrow from one number and pay it back to another." As late as five years ago, when the author was explaining how to teach the additive method, a teacher in the audience rose and said, "Your carrying to the lower number is in reality borrowing one above and paying it back to the number below it. This is unethical and I shall never use it."[43]

In the eighteenth and nineteenth centuries many arithmetics in this country used the equal additions method in one or more of the three phases of subtraction. Nathan Daboll's[44] famous text, which appeared in 1797 and ran through many editions, used this method of subtraction in mixed numbers and denominate numbers.

H. N. Robinson's popular text of the nineteenth century, surviving many editions, used the equal additions method in all three phases of subtraction.[44a]

In France after the beginning of the nineteenth century the equal additions method seemed to be the established custom. Cirodde,[45] in 1858, used it in all three phases of subtraction. A textbook by Jules Tannery[46] contains the equal additions method in whole number subtraction, but in mixed number subtraction the

[42] Stone, J. C. *How We Subtract*, p. 23. Benj. H. Sanborn & Co., Chicago, 1926.
[43] *Ibid.*, p. 46.
[44] Daboll, Nathan. *School Master's Assistant*. Mack, Andrus and Woodruff, Ithaca, N. Y. 1797 (1837 ed.).
[44a] Robinson, H. N. *Practical Arithmetic*. Ivison, Blakeman, Taylor and Company, New York, 1881.
[45] Cirodde, P. L. *Leçons d'Arithmétique*, p. 130. L. Hachette et Cie, Paris, 1858.
[46] Tannery, Jules, *Leçons d'Arithmétique et Pratique*. Colin et Cie, Paris, 1900.

author goes back to the medieval method of changing mixed numbers to improper fractions and then to common denominators.

As an example of this cumbersome method of subtracting mixed numbers, the following explanation from Humphrey Baker may be illuminating:

If you will substract whole numbers and broken from whole and broken; as thus, if you will substract 9 1/4, from 20 1/2, you must reduce 9 1/4, into fourths, and likewise the 20 1/2, into halfes by the first reduction, and you shall finde 37/4 for the 9 1/4; and 41/2 into one denomination: according unto the firste reduction, and you shall finde 74/8 for the 37/4; and 164/8, for the 41/2, then abate the numerator of 74/8., which is 74. from 164 which is the numerator of 164/8, and there remaineth 90/8, then divide 90 by 8, and thereof cometh 11 1/4, which is the remaine of this substraction.[47]

The example is set forth as follows:

$$\frac{37}{9\frac{1}{4}} \qquad \frac{41}{20\frac{1}{2}} \qquad \frac{74}{} \qquad \frac{164}{}$$

$$ 90 \qquad \frac{41}{2}$$

$$\frac{37}{4}$$

$$\begin{array}{c} 164 \\ 74 \\ \hline 90 \end{array} \qquad \begin{array}{l} 8 \\ \cancel{1}\cancel{2} \\ \cancel{9}\cancel{0}\ (11\frac{1}{4}) \\ \underline{\cancel{8}\cancel{8}} \end{array}$$

This method in mixed number subtraction was found as late as 1850 in an arithmetic by McCurdy[48] in this country and as late as 1883 in one by Spickernell[49] of London.

As an illustration of the equal additions method in mixed number subtraction the following example is given from a text by French authors.[50]

Let it be required to solve: (3 1/3 − 2 4/5)
Let us add one or 3/3 to 1/3: the upper fractional number becomes 3 4/3. The larger number having been increased by one, it is necessary to increase the number to be subtracted by the same amount. The subtraction is indi-

[47] Baker, Humphrey. *The Wellspring of Sciences*, pp. 67, 68. London, 1568 (1591 ed.).

[48] McCurdy, D. *New American Order of Arithmetic*. Armstrong & Berry, Baltimore, 1850.

[49] Spickernell, G. E. *An Explanatory Arithmetic*, pp. 51, 52. Griffin & Co., London, 1883.

[50] Lemoyne-Aymard. *La Théorie Arithmétique*, p. 213. H. Paulin et Cie. Paris, 1910.

cated thus (3 4/3) − (3 4/5) and one proceeds as it has been indicated in n.873.[51] (n.873 refers to first method, which was used on the fractional parts only.)

A later method book, by Stuyvaert,[52] gives only the one method of equal additions in all of its subtractions, both whole numbers and mixed. Naturally, denominate number subtraction was not treated in a country where the metric system is used. Many of the later French books made much of propositions and theorems pertaining to subtraction. Stuyvaert's book gave nine theorems on subtraction alone. Theorem number 17 pertains to the subtraction principle and is here quoted to show one theorem typical of many in other French books:

A difference is not changed when one adds the same number to its two terms.[53]

Thus the equal additions method seems to be well established as the prevailing procedure in France,[54] although other methods are also being used.

The equal additions method has also been extensively used in England. Augustus De Morgan in his *Elements of Arithmetic* says, in explaining the equal additions method which he used in all three phases of subtraction:

The process of subtraction depends upon these two principles. The difference of two numbers is not altered by adding a number to the first if you add the same number to the second; or by subtracting a number from the first, if you subtract the same number from the second. Conceive two baskets with pebbles in them, in the first of which are 100 pebbles more than in the second. If I put 50 more pebbles into each of them, there are still only 100 more in the first than in the second, and the same if I take 50 from each. Therefore in finding the difference of two numbers if it should be convenient,

[51] 2e Méthode. Soit à effectuer (3 1/3) − (2 4/5)
Ajoutons 1 unité ou 3/3 à 1/3; le nombre fractionnaire le plus élevé devient 3 4/3. Le plus grand nombre ayant été augmenté d'une unité, il faut augmenter le nombre à retrancher de la même quantité. La soustraction s'indique ainsi (3 4/3) − (3 4/5) et on procède comme il a été indiqué n. 873."
[52] Stuyvaert, M. *Les Nombres Positifs*, p. 9. Van Goethem & Cie, Gand, Belgium, 1912.
[53] "Une différence n'est pas altérée quand on additione une même nombre à ses deux termes."
[54] In very recent years the Austrian method is being used there.

I may add any number I please to both of them, because, though I alter the numbers themselves by so doing, I do not alter their difference.[55]

The equal additions method prevailed in the United States in the eighteenth and early nineteenth centuries, as has already been mentioned. It is still being used to some extent, but since 1850 it has been largely superseded by the decomposition method sponsored by Colburn in his famous *Arithmetic*.

THE AUSTRIAN METHOD

Possibly the latest method from the point of historical use is the Austrian, described by David Eugene Smith as the addition method:

The addition method, familiar in "making change," is possibly the most rapid method if taught from the first. To subtract 87 from 243 the computer says: "7 and 6 are 13; 9 and 5 are 14; 1 and 1 are 2"; or else he says: "7 and 6 are 13; 8 and 5 are 13; 0 and 1 are 1," the former being the better. The method was suggested by Buteo (1559) and probably by various other early writers, but it never found much favor among arithmeticians until the 19th century. It has been called the Austrian method, because it was brought to the attention of German writers by Kuckuck (1874), who learned of it through the Austrian arithmetics of Mŏcnik (1848) and Josef Salomon (1849).[56]

$$\begin{array}{r} 243 \\ 87 \\ \hline 156 \end{array}$$

In this example Smith gives two explanations. The first "7 and 6 are 13; 9 and 5 are 14; 1 and 1 are 2" is the Austrian, as used by Stone* and by Strayer-Upton.** The second "7 and 6 are 13; 8 and 5 are 13; 0 and 1 are 1" is the method described in the introduction to this study (page 3) as the additive variety of the decomposition method. If we call the first method Austrian, what shall we call the second? If we call them both additive methods —as in a sense they are—we are confusing terms by giving the name "additive" to a method that decreases the minuend figure. Though the second method is rarely taught at the present time, it nevertheless is used to a small extent. It is evidence of con-

---

[55] De Morgan, Augustus. *Elements of Arithmetic*, p. 21.   J. Taylor, London, 1835.

[56] Smith, David Eugene. *History of Mathematics*, Vol. II, p. 101. Ginn and Company, Boston, 1925.

* Stone, J. C. *The Stone Arithmetic, Third Year*, p. 39. Benj. H. Sanborn & Co., Chicago, 1929.

** Strayer, G. D. and Upton, C. B. *Strayer-Upton Arithmetics. Lower Grades*, p. 64. American Book Co., New York, 1928.

fusion again caused by combining the additive language with the decomposition technique.

Karpinski says concerning the Austrian method:

> The strictly addition procedure in subtraction is mentioned in the *Handbuch der Mathematik* by Bittner, published at Prague in 1821; the method is explained in Salomon's *Lehrbuch der Arithmetik und Algebra*, Vienna, 1849. In America this has been known as the Austrian method, and its use is recommended to primary teachers in the courses of study of several large school systems.

826
483
———
343

Think of the number which added to 483 will give 826; 3 added to 3 gives 6; 4 added to 8 gives 12; write down 4 and mentally carry 1 to the next 4, making 5; 3 added to 5 makes 8. 343 is the number which added to 483 gives 826.[57]

The earliest use of the Austrian method in this country that could be discovered by the writer was the so-called computer's method found in a textbook of 1902. The following illustration shows the peculiar way in which the example was written.

From 94,275 take 67,492.

94275
67492
———
26783

Thus: 2 and 3 are 5; 9 and 8 are 17, carry 1 to 4 as in addition, making it 5, 5 and 7 are 12; carry 1 to 7 making it 8; 8 and 6 are 14; carry 1 to 6 making it 7; 7 and 2 are 9.[58]

The explanation of the example is typical of the Austrian method, though the form is not. The form is perhaps due to the idea that, since it is an addition method, it had to have the form of addition; and hence became inverted because it had to satisfy the subtraction idea. This form of inverted addition is no longer being used.

Stone says concerning the Austrian method:

> While the use of the additive method was advocated by our leading educators, but little was written to show how to teach it. It was first advocated at a time when the decomposition method was almost universally used, and teachers attempted to "explain" *this* method when any number in any column was less than the one below it by decomposing the next left-hand number of the minuend. At least one text-book gave the same method. Thus, in the following exercise:

[57] Karpinski, L. C. *History of Arithmetic*, p. 105. Rand McNally and Co., Chicago, 1925.

[58] McClellan and Ames. *Public School Arithmetic*, p. 36. Macmillan Co., New York, 1902.

The pupil was told to take 1 of the 3 and think, "8 and 7 are 15." Then take 1 of the 4 and think, "6 and 6 are 12." And then think, "1 and 2 are 3."

435
168
267

This was a very common method of procedure when the method was first introduced, and the writer has known many large city systems trying to teach it by this method within the last few years. This procedure has no advantages over the take-away method using decomposition, unless it is the saving of time in learning the tables. The method thus taught becomes complicated and requires the increased attention of decomposition. However, since no method can ever displace an old one unless it has very decided advantages in its favor, the additive method has never succeeded when retaining the decomposition features.[59]

The only case which the writer found of the use of the Austrian method in mixed number subtraction was in a text by Halsted. The method does not seem well adapted to this kind of subtraction, as the reader may judge from the following example:

13 1/4
7 3/4
5 2/4

To subtract 7 + 3/4 from 13 + 1/4, that is to evaluate 13 1/4 − 7 3/4, think 3/4 and two-fourths make 5/4, carry 1; 8 and 5 make 13.[60]

The method recommended by Stone for subtraction of mixed numbers is really not Austrian but complementary. Note his example:

Subtract 3⅝ from 5⅛

5⅛
3⅝
1½*

The child thinks, "⅝ and ⅜ = 1; ⅜ + ⅛ = ½." Write ½, carry 1. Then, "4 and 1 are 5." Write 1.[61]

To be consistent with the Austrian method of whole numbers the above example would read thus:

5/8 and 4/8 = 9/8; write 4/8 or 1/2, 4 and 1 are 5. Write 1.

Current Methods in Europe

From recent communications with leaders in arithmetic in England, France, and Germany with regard to the prevalent current methods of subtraction in these countries, the writer obtained the information given below.

* Answer supplied by the writer.

59 Stone, J. C. *How We Subtract*, p. 69. Benj. H. Sanborn & Co., Chicago, 1926.

60 Halsted, G. B. *Foundations and Technique of Arithmetic*, p. 61. Open Court Publishing Co., Chicago, 1912.

61 Stone, *op. cit.*, p. 92.

The question was asked how this example in subtraction:

42
27

was taught in their schools; whether by

(a) the decomposition method: 7 from 12, 2 from 3, or by
(b) the equal additions method: 7 from 12, 3 from 4, or by
(c) the Austrian method: 7 and ? are 12, 3 and ? are 4, or by
(d) the complementary method: 7 from 10, 3 and 2 are 5, 3 from 4.

The question was also asked which of the above methods was used in mixed number and denominate number subtraction. The following replies and comments were received.

Dr. Herbert R. Hamley of the University of London wrote under date of January 29, 1935:

With regard to your questions all the methods you describe are to be found in English schools, for there are good and bad methods just as there are good and bad schools. On the whole I should say that the most popular methods in this country are the Austrian and the equal additions methods. We regard the Austrian method as a method of equal additions. The complementary method is seldom used, and is frowned upon by all our authorities. The decomposition method is occasionally found among older teachers. On the whole I think you would find the following jargon in most of our elementary schools:

42	7 to make 2—cannot (be done).
27	7 to make 12—5. Put down 5.
15	Add 1 to 2, making 3. 3 to make 4—1. Put down 1.

8¼	½ to make ¼—cannot (be done).
3½	½ to make 1¼—¾. Put down ¾.
4¾	Add 1 to 3, making 4. 4 to make 8—4. Put down 4.

Pounds	Shillings	Pence	
9	11	3	8 pence to make 3 pence—cannot (be done). 8 pence to make one shilling and 3 pence—7
6	15	8	pence. Put down 7.
			Add one shilling to 15 shillings, making 16
2	15	7	shillings. 16 shillings to make 11 shillings—cannot (be done). 16 shillings to make £1.11 shillings—15 shillings. Put down 15. Add £1 to £6, making £7. £7 to make £9—2. Put down 2.

Professor Albert Chatelet, rector of the Academy of Lille, France, sent copies of four textbooks in arithmetic and wrote under date of February 6, 1935, as follows: (Words in parentheses are the writer's)

For the subtraction of whole numbers the most usual method is method (b) (the equal additions method).

42   You cannot subtract 7 from 2; 12 less 7 makes 5; I write down 5 and
27   carry 1; 2 and 1 make 3, 4 minus 3 make 1; I write down 1.
——
15   Many teachers, however, use method (c) (the Austrian method) and I am inclined personally to use this.

7 cannot be subtracted from 2, but from 12; 7 and 5 makes 12; I write down 5 and carry 1; 2 plus 1 makes 3, 3 plus 1 makes 4; I write down 1.

The complementary method (d) is rarely used and I recommend it only for mental arithmetic in case of numbers of one figure. In the example quoted, I should say: the subtraction causes a carry-over. I take away the higher ten: 42 minus 30 = 12. I add the complement of 7. 12 and 3 = 15.

Method (a) of decomposing is rarely used save in mental arithmetic and then in inverse order, tens and unities.

$$42 \text{ less } 20 = 22$$
$$22 \text{ less } 7 = 15$$

Compound numbers, $9\frac{1}{2}$, $6\frac{3}{4}$, are but little used now in France; they would be written 9,5; 6,75. The subtraction of these decimal figures is done in the same way as that of whole numbers.

Complex numbers[62] are also little used. You will find, nevertheless, four lessons on these numbers. Thus subtraction by the method of addition is indicated.

Experiments carried out with a number of pupils show that they easily assimilate any method. Nevertheless, teachers in the habit of using method (b) (equal additions), who had but tentatively used method (c) (Austrian), have the impression that the pupils used it (Austrian) with greater facility. The advantage I see in it is to do away with the distinction between the addition and subtraction tables.

Dr. William Lietzman, director of the State Oberrealschule of Göttingen, wrote on January 14, 1936, as follows:

In German schools at the present time are used:
(1) the so-called North-German method of subtraction in the form:

42   7 from 12 is 5
27   2 from 3 is 1
——
15

(2) the so-called Austrian or South-German method:

7 and 5 is 12
3 and 1 is 4

[62] Meaning denominate numbers.

also in the following form:

from 7 to 12 is 5,
from 3 to 4 is 1,

with which the number to be written in the answer stands *at the end.*

The equal additions method is not commonly used.

In the case of fractions and denominate numbers the method employed is guided by the method used at other times.

The limited study given to the history of methods of subtraction reveals that the complementary, the decomposition, and the equal additions methods were in use long before the art of printing was invented. We owe the use of the term "borrow" to the mechanical device known as the abacus.

It is interesting to note that the equal additions method was not found in any German text examined, the decomposition method being chiefly used in that country. On the other hand, the decomposition method was not found in any French book examined, published later than 1820, the equal additions method being the prevalent method in France. The complementary and equal additions were the outstanding methods in Italy and England. The Austrian method was found in very recent English texts but did not appear in any Italian books examined.

In the United States all methods have been used in a sort of chronological progression as shown by C—E—D—A. Very recently, however, the complementary method is appearing again in this country,[63] though perhaps even the authors using it are not aware of its origin.

The question of procedure in subtraction is, today, more unsettled in this country than abroad.

[63] See treatment of mixed and denomination numbers in Stone-Mills, *Unit Mastery Arithmetics, Intermediate Book*, pp. 277 and 439. Benj. H. Sanborn & Co., Chicago, 1932.

III

PAST INVESTIGATIONS RELATING TO METHODS OF SUBTRACTION

Both England and the United States have made rather extensive investigations of methods in subtraction. In England the issue was between equal additions and decomposition. In 1914 Ballard gave an arithmetic test to 9,176 boys and 9,502 girls to establish norms. The timing and scoring were objectively done. Ballard states the results as follows:

> Suffice it to say that for every age the E. A. (equal additions) children of both sexes were found to work subtraction more expeditiously than the D. (decomposition) children. And on the whole the number of errors is less.[1]

Ballard's study was followed by two others, one in 1918 by W. W. McClelland, and another in 1920 by W. H. Winch.

McClelland tested 63 pupils with the equal additions method and 80 with the decomposition method. The duration of the tests for each group was ten minutes. In that time the equal additions group subtracted 161 columns with 6.7% errors and the decomposition group subtracted 134 columns with 9.7% errors. Both groups were later given additional practice over a long period of time. McClelland concluded by saying:

> The method of equal additions appears superior in speed, accuracy, and adaptability to new conditions, while the method of decomposition is superior in speed after long practice.[2]

The investigation by Winch was somewhat different. Thirty-eight girls twelve years of age, all having been taught the decomposition method, were divided into two equal ability groups.

---

[1] Ballard, P. B. "Norms of Performance in the Fundamental Processes of Arithmetic." *Journal of Experimental Pedagogy*, Vol. Il, pp. 396-405, December, 1914; also Vol. III, pp. 9-20, March, 1915.

[2] McClelland, W. W. "An Experimental Study of the Different Methods of Subtraction." *Journal of Experimental Pedagogy*, Vol. IV, pp. 293-299, December, 1918.

For the two succeeding months one group continued with the decomposition method, while the other group were taught the equal additions method, which was new to them, and were given practice in it. All teaching and testing of both groups were done by the same teacher. Winch's conclusions are here quoted:

(1) The method of equal additions in subtraction taught to children late in school life, who have hitherto worked by decomposition, produces results in a few weeks equal on the whole, and superior in the weaker children, to those produced by the method of decomposition. (2) The amount of the gain involved does not justify a change at this late period of a child's school career.[3]

Additional weight would seem to attach to the first of these conclusions in view of the fact that the second group were heavily handicapped first in unlearning the familiar decomposition method and then in learning and practicing a new method in the same length of time as was spent by the first group in improving their own familiar method of decomposition.

The results led Winch to try again later with pupils of an average age of eight and one-half years, most of them just learning to subtract. He says the results were more marked than before and states that in the case of both inferior and superior children, "The method of equal additions shows to decided advantage with young children in accuracy and rapidity." [4]

No further experiments are reported from England. Apparently the case is closed in that country in favor of the equal additions method.

In the United States the interest was at first directed to the relative merits of additive and subtractive (take-away) methods.

Mead and Sears report the first investigation in 1916.[5] The difference found was very small. The report concludes by saying that if allowance is made for the slight initial superiority of the additive group, the outcome of the final tests favors the take-away group three times out of five; that if test 5 is ignored (because of unreliability), the outcome favors neither group.

[3] Winch, W. H. "Equal Additions versus Decomposition in Teaching Subtraction." *Journal of Experimental Pedagogy*, Vol. V, pp. 207-220, 261-270, June, 1920.

[4] *Ibid.*

[5] Mead, C. D. and Sears, I. "Additive Subtraction and Multiplicative Division Tested." *Journal of Educational Psychology*, Vol. VII, pp. 261-270, May, 1916.

In 1919 J. S. Taylor[6] made a study of the methods of subtraction actually used in two districts of New York. Superintendent Maxwell had specified that subtraction should be taught "by the addition process" in the schools of the city. After six years Taylor found that the methods used were distributed as follows: decomposition method, 21.8%; equal additions method, 40.6%; Austrian method, 37.6%. He also found that the number of pupils using the Austrian method dropped from 52.7% in the fourth grade to 21.2% in the sixth grade.

Taylor's conclusion that the Austrian method was a failure in New York may not be correct because, no doubt, there were many teachers who failed to follow out the prescription and taught other methods besides the Austrian. This report proves nothing with regard to the efficiency of either method. Some interesting data might have been had if the accuracy results obtained from this study of 11,368 pupils using three different methods had been saved.

W. W. Beatty[7] in 1920 conducted a similar experiment in California, where the Austrian method had been taught. He tested 175 upper grade pupils and found that 115 used the "borrow" method. He gives no evidence of having checked to discover whether these 115 pupils were the same 115 who had actually been taught the Austrian method. Transfers and changes in teacher personnel may have accounted for results in many of the 115 cases. The results are not convincing for the difference in accuracy favors the Austrian group by 81.7% to 79.3%, while the difference in rate favors the borrowing group by 9.2 to 8.2 examples done.

In 1927 B. R. Buckingham[8] made a similar study of additive versus take-away methods in subtraction. Seven schools were used, in which pupils were paired into comparable groups on the basis of intelligence. Length and number of class periods were the same for both groups, but one group was taught the take-away method and the other the additive. The results from the

---

[6] Taylor, J. S. "Subtraction by the Addition Process." *Elementary School Journal*, Vol. XX, pp. 203-207, November, 1919.

[7] Beatty, W. W. "The Additive versus the Borrowing Method of Subtraction." *Elementary School Journal*, Vol. XXI, pp. 198-200, November, 1920.

[8] Buckingham, B. R. "The Additive versus the Take-Away Method of Teaching the Subtraction Facts." *Educational Research Bulletin*, No. 6, pp. 265-269. Ohio State University, 1927.

final tests showed that in six of the seven schools the take-away method was superior to the additive. The probable error of the difference between means, however, was so large as greatly to diminish the reliability of the results. The difference in the results of the tests in the six groups favoring the take-away method ranged from **7.2** to **13.5**, while the probable error of the difference ranged from **3.8** to **9.8**. In the one school favoring the additive method the difference in the results of the test was **8.3** and the probable error of the difference was **4.0**.

A study by W. F. Roantree in 1924 compares results of the Austrian method with those of the decomposition (called by him the Italian) method in earlier grades 2, 3, and 4. The results of the first test, given at the end of the second grade to 346 children in five New York City schools, are not reliable because we do not know that the pupils had had time to learn the compound subtraction technique. A second attempt was made by giving the same test in the Model School of the New York Training School for Teachers and in a neighboring school during the last week of every term. The results from the 2B classes, the only classes for which the data were reliable, showed that the Austrian method was superior. In the same test, given to grade 3 and grade 4B, the comparative results showed no significant difference.

The two conclusions which were drawn from this study of Roantree were as follows:

First, the superiority of the Austrian method in the work of the second year indicates that this method is more easily learned than is the Italian method.
Second, it probably makes little difference, if any, which method one uses, assuming that the method used has been thoroughly mechanized.[9]

In 1923 the present writer[10] became interested in this subject at the time he first taught methods in arithmetic at the Chicago Normal College. His own method was decomposition, which he considered better than other methods. To get some information and corroboration in this matter he gave the Courtis Lesson Card No. 33 to each of his classes for two semesters, and carefully

[9] Roantree, W. F. "The Question of Method in Subtraction." *Mathematics Teacher*, XVII, pp. 110-116, February, 1924.
[10] Johnson, J. T. "The Merits of Different Methods of Subtraction." *Journal of Educational Research*, Vol. X, No. 4, pp. 279-290, November, 1924.

checked time and accuracy for each student. The Courtis card contains seventeen examples like the one below. Each student was asked just how she subtracted each step in the example $\frac{137017}{70719}$.

Four methods of subtraction were found: subtractive-borrowing, subtractive-carrying, additive-borrowing, additive-carrying. To the surprise of the instructor, those who used the carrying methods came out ahead in all of the eleven sections except one, and still more surprising, not only were their results more accurate but they arrived at them in less time than did those who used the "borrow" or decomposition method. Since this was more than could be accounted for by chance, the writer's interest was focused at once on the problem, and he has been collecting data ever since. The results of this first study, based on but 277 cases, proved nothing, but they did show a difference both in accuracy and in time in favor of the methods that increased the subtrahend. The results were as follows:

	Accuracy (%)	Time
I. Decomposition	92.5	2 : 55
II. Additive borrowing	92.4	2 : 40
III. Equal additions	96.5	2 : 30
IV. Austrian	96.4	2 : 10

Combining methods I and II, those which diminished minuends, the results were: accuracy 92.45% and time 2:50. Combining methods III and IV, those which increased the subtrahends, the results were: accuracy 96.45% and time 2:20. This is a difference of 4% in accuracy and at the same time 30 sec. of time, both in favor of the group that increased the subtrahend. Combining methods I and III, the subtractive methods, accuracy results were 94.5% and time, 2:43. Combining methods II and IV, the additive methods, accuracy results were 94.5% and time, 2:30. Thus there resulted no difference in accuracy between the additive and subtractive methods, and but a slight difference in time in favor of the additive.

Next, Osburn[11] in 1927 made a study corroborating former findings concerning the decomposition and equal additions methods. He classified subtraction into three main methods: additive, take-

[11] Osburn, W. J. "How Shall We Subtract?" *Journal of Educational Research*, Vol. XVI, pp. 237-246, November, 1927.

away, and complementary. He then subdivided these three into two each, on the basis of whether the minuend is decreased or the subtrahend increased. He then subdivided each of these six methods into two each on the basis of direction of the subtraction, up or down, thus distinguishing twelve methods.

Osburn gave a test of twelve examples such as this one:
$$\frac{15210}{8972}$$

to twenty-three school systems including 1,414 pupils in Wisconsin. On the basis of his findings he divided the results into the classifications below.

	Arithmetic Mean of Errors	Standard Deviation	Number of Cases
Take-away			
Decomposition			
Upward (A).................	2.7 ± 0.14	4.1	883
Downward (B)..............	2.6 ± 0.54	3.2	34
Equal additions			
Upward (C).................	1.9 ± 0.20	2.9	209
Downward (D)..............	1.6 ± 0.26	1.4	29
Additive			
Decomposition			
Upward (E).................	2.8 ± 0.49	3.6	54
Equal additions			
Upward (F).................	2.1 ± 0.25	3.2	148

Since the downward methods had few users, Osburn combined them, making the following classifications:

	Mean Errors	S. D.	N
Take-away decomposition			
(A and B)...................	2.7 ± 0.13	4.0	917
Take-away equal additions			
(C and D)..................	1.8 ± 0.18	2.7	238
Additive, both forms			
(E and F)..................	2.3 ± 0.25	3.4	202

Osburn points out that the difference between means in decomposition and equal additions is statistically significant.

Following upon his study of 1923 already recorded, the results of which, favoring the E method, had been quite contrary to his expectations, the present writer began a more extended investigation of the same problem. During the years 1924 to 1930, he collected data from each of his Normal College classes and from grades 5, 6, and 8 in the Chicago public schools, until he had results from 464 grade pupils, 140 Normal College students, and 89 university graduates, a total of 693 persons. Tests revealed that of these, 355 used the equal additions technique and 338 the decomposition technique. Though among the former there were some who used the Austrian method, he did not differentiate them from the rest of the 355, since all in that group increased the subtrahend figure. This differentiation he based on a long series of observations and tests, formal and informal, which had led him to the conclusion that the same person may and frequently does use both subtractive and additive language, often changing from one to the other form, according to the difficulty of the example or, perhaps, because of a habit of checking his subtraction. In a very simple operation, for instance, as in subtracting 6 from 8, most people would say to themselves simply "6 from 8," automatically condensing the mental process. A degree of introspection might bring, "6 lacks but 2 of being 8, therefore 6 from 8 is 2," or "It takes 2 more with 6 to make 8," this latter form of thinking being definitely additive. In short, many people are not clear about what goes on in their minds when they subtract. Hence, it seemed wise to make a division of methods, not on so unclear a basis as additive or subtractive language, but rather on whether the minuend figure was decreased or the subtrahend figure increased, these two procedures not having been used by the same person.

In an effort to isolate the technique to be measured, the writer made two tests. One, which might be considered the control test, consisted of the 100 simple facts only; the other was made up of 30 examples in subtraction where the technique to be measured would have to be employed.

As both tests were taken by each of the subjects and the excess of time and errors on the second test over the first test was measured, the personal equation elements were eliminated each time, leaving but the extra time taken for the decomposition technique or the equal additions technique. The personal equation, if it may be so called, consisted of several elements, the chief of which were native intelligence, subtracting ability, previous practice in subtraction, physical condition, emotional condition, and, most important, the reaction time, which varies with individuals. All of these elements were operative in both tests; hence their effects were eliminated in subtracting the time and number of errors in the first test from the time and numbers of errors in the second test, leaving only in the subtraction situation the time and number of errors caused by the technique of decomposition or of equal additions. The results of the author's two tests were as follows:

	Decomposition	Equal Additions
Mean accuracy in the 100 subtraction facts.	99.06 ± 1.3	99.08 ± 1.2
Mean accuracy on the 30 examples........	87.19 ± 8.5	90.19 ± 7.5
Difference, or loss in accuracy, due to the technique used	11.87	8.89
Median time on the 100 subtraction facts..	3:21 ± 1:92	3:43 ± 1:02
Median time on the 30 examples..........	6:32 ± 2:01	5:58 ± 1:12
Difference, or loss in time, due to the technique used.........................	3:11	2:15[12]

Computing the probable error of the difference between the means in accuracy from the above results we obtain,

$$\text{P.E. (diff. } m_D - m_E) = \sqrt{\text{P.E.}m_D^2 + \text{P.E.}m_E^2} = \sqrt{\left(\frac{8.54}{\sqrt{338}}\right)^2 + \left(\frac{7.45}{\sqrt{355}}\right)^2} = .6$$

Actual difference between means $= (90.19 - 87.19) = 3.0$. The critical ratio $= \frac{3.0}{.6} = 5$, which indicates a significant difference in favor of the E method. When the difference in the gains in time on the 30 examples is taken into account, the results become more significant. This difference in the extra time for the 30 examples was 56 seconds in favor of the E method. In terms of per cent this is $\frac{0:56}{3:11} = \frac{56}{191} = 29.3\%$.

In the following year, 1932, the question was discussed in a Canadian journal by H. Bowers.[13] In two successive numbers of this journal Bowers comments on the confusion of so many names

[12] Johnson, J. T. "The Efficiency of Decomposition Compared with That of Equal Additions as a Technique in Subtraction of Whole Numbers." *Mathematics Teacher*, Vol. XXIV, pp. 5-13, January, 1931.

[13] Bowers, H. "Methods of Subtracting." *The School*, Vol. XXI, pp. 7-12, 105-108, September, October, 1932.

and methods used in subtraction. He makes his own contribution by adding one more method and new meaning for the words process, method, and procedure. The method is that of "delayed compen-

53
26
—
27

sation," illustrated by the example at the left thus: "6 from 13, 7; 2 from 5, 3; 1 from 3, 2." He says that procedure should be synonymous with both method and process, but method and process are distinct in meaning. He names the three methods of subtraction as complementary, take-away, and additive, and the two processes which may be used with any one of the methods, as equal additions and decomposition. He comments favorably on the complementary method and shows its advantages in mixed number and in denominate number subtraction.

Bowers then goes on with the arguments for and against the processes of decomposition and equal additions. He lists the arguments usually given for and against these methods. The main argument for the decomposition method, he says, is the fact that it can be explained better. He answers this argument by saying that the fact that a procedure is easy to explain is hardly a valid reason for its retention. He then cites Ballard's reference to Stephenson's first steam engine as being easier to understand than a Rolls-Royce car, but having on that account no advantage over the latter. He expresses this thought in his own words when he says,

When we consider that subtraction, however carried out, becomes sooner or later a mechanical procedure, it will be agreed that by itself an initial advantage in the matter of rationalization is of relatively small importance.[14]

His own arguments are distinctly in favor of equal additions as against decomposition; and he backs up his argument by sound evidence, both logical and experimental. He mentions the close agreement between Ballard, McClelland, Winch, and Osburn on the inferiority of the decomposition processes to the equal additions process, and confirms their conclusions further by giving the results of an experiment not previously reported in this study. It was made in 1925 by one H. M. Inspector, in the County of Gloucestershire. To quote from Bowers:

[14] Bowers, H. "Methods of Subtracting." *The School*, Vol. XXI, p. 11, September, 1932.

In the latter case, of the 118 schools tested it happened by pure coincidence that half had adopted the one process and half the other. The averages obtained by the two types of methods were:

	Schools	Average (number of examples) in 3 Minutes
Decomposition..........	59	2.3
Equal additions........	59	3.1

It is hardly necessary to say that in comparing two processes the method used would be kept constant.[15]

He quotes the following from Ballard:

I have watched children near the top of the senior school subtracting, say 17 from 30,000, and heard them mutter: "7 from 0, cannot, go next door. Nothing there; go next door. Nothing there; go next door." And they continue this Mother Hubbard business till they get to the 3. Then they put a little 2 over the 3, 9 over each of the next three noughts, and 10 over the last nought.[16]

He next goes into the controversy of additive and take-away subtraction. The method of making change is brought up as not a parallel to subtraction, for the difference is not found when making change. Thorndike makes this clear in his *New Methods in Arithmetic,* where he says, "Making change is one thing. . . . Subtraction is a very different thing." [17]

Bowers emphasizes the importance of considering subtraction methods in the other two phases, those of mixed numbers and denominate numbers.

The question has not reached final settlement. In March, 1934, two years after the present investigation was begun, Wilson in an article in the *Journal of Educational Research* [18] gave a digest of Blanch M. Allen's master's thesis: "Subtraction: Current Methods of Subtraction in the United States," a study based on data from 23 departments of education, 162 cities, 215 training schools, normal schools, and teachers colleges. Wilson tabulates the results of Allen's findings and says the take-away method is used three times as often as the additive; borrowing is used two and one-half times as often as equal additions; upward is six times as often as

---

[15] *Ibid.,* p. 106.
[16] *Ibid.,* p. 12.
[17] Thorndike, E. L. *New Methods in Arithmetic,* pp. 216-217. Rand McNally and Co., Chicago, 1921.
[18] Wilson, G. M. "For 100% Subtraction, What Method? A New Approach." *Journal of Educational Research,* Vol. XXVII, pp. 503-508, March, 1934.

downward. Take-away borrowing upward, then, is the method he recommends, since that procedure got 48 per cent of the total votes. No other received one-third as many.

Wilson comes down to two methods, after discarding the Austrian and the complementary methods because so few people use them. These two methods are the decomposition and equal additions methods. Of these two, Wilson recommends the decomposition method. He gives as his reason the fact that it is used $2\frac{1}{2}$ times as often as the equal additions method.

The assumption upon which Wilson bases his argument is, of course, in turn based upon a status quo philosophy, viz., that whatever is the current practice is the best practice. The present writer is interested to know which is the better method regardless of whether it is the current method or not.

Another study in the form of a master's thesis [19] is worth summarizing in that its approach and method are distinctly different from those of the other studies reported.

Margaret Daniel, the author of this thesis, secured second grade pupils before any subtraction whatever had been taught them. During the fall term one group of 146 were taught the simple subtraction facts by the take-away method and another group of 156 were taught the subtraction facts by the additive method using "and." During the following spring term when subtraction with minuend figure smaller than the subtrahend figure was taken up in examples like $\frac{52}{38}$ each of these groups was divided again and about one half of each group was taught the decomposition technique and the other half the equal additions technique.

It is evident that four methods resulted from these two divided groups. The groups, as named by Miss Daniel, with the number in each follow:

	N
I. Take-away equal additions	78
II. Take-away decomposition	68
III. Additive equal additions	77
IV. Additive decomposition	79

The four groups were equated (1) in mental ability, using the

[19] Daniel, Margaret D., "Comparative Merits of Four Methods of Teaching Subtraction." Unpublished Master's Thesis, University of California, August, 1931.

Haggerty Intelligence Test, Delta I, and (2) in adding ability, using the Compass Diagnostic Test in Arithmetic, Test I.

The ranks of the four groups in subtraction based on the totals for the fall and spring terms were:

	Groups			
	I	II	III	IV
Fall term.............	3	1	2	4
Spring term..........	1	4	2	3

Using Garrett's formula for the reliability of the difference between means in terms of P.E., Group II (the group that had used the take-away method on the simple subtraction facts) were superior, at the end of the fall term, to Group I (the group that had also used the take-away method on the simple facts) by $3.5\text{P.E.}_{(\text{diff.})}$.

At the end of the spring term Group I (the group that had used the take-away method on the simple facts but later had used the equal additions technique on the subtraction examples) were superior to Group II (the group that had also used the take-away method on the simple facts but had later used the decomposition technique on the examples) by $3.1\text{P.E.}_{(\text{diff.})}$. In non-statistical terms, a difference between means equal to 3.1 times the probable error of the difference indicates that there are 98 chances in 100 that there is a true difference in the same direction—in this case in favor of the equal additions technique.

The differences between Groups III and IV were not significant. Besides, the methods used by these groups (additive language with equal additions technique and additive language with decomposition technique) are seldom if ever taught. Hence value is placed on results of methods I and II only.

The technique used in Group III was not the Austrian technique.

The study seems to show that with pupils who had been taught the take-away method on the simple subtraction facts, the equal additions technique later produces results superior to those produced by the decomposition technique.

Nothing is shown with regard to those pupils who had been taught the additive method on the simple subtraction facts.

Out of the fourteen studies in subtraction here reviewed, ten are experimental studies and four are discussions either on relative merits or on permanence of the methods. Of the ten experimental studies, one is unreliable and two deal with the additive *vs.* the subtractive method in simple subtraction. This leaves one that deals with the Austrian *vs.* the decomposition method and six that deal with equal additions *vs.* decomposition. These seven experiments by Roantree, Ballard, McClelland, Winch, Johnson (second experiment), Osburn, and Daniel are all fairly reliably reported, and all of them show that the decomposition method is inferior to a method that increases the subtrahend—either the equal additions method or the Austrian.

IV

METHOD OF PROCEDURE IN THE PRESENT INVESTIGATION

THE aim of research should be the establishment of an empirical law. The method of research should be that of science.

The methods of science are three: the normative, the historical, and the experimental. John Stuart Mill[1] gave five canons of experimentation: the method of agreement, the method of difference, the joint method of agreement and difference, the method of non-comitant variation, and the method of residue. He states the method of difference as follows:

If an instance in which the phenomenon under investigation occurs, and an instance in which it does not occur, have every circumstance save one in common, that one occurring only in the former; the circumstance alone in which the two instances differ, is the effect, or cause, or a necessary part of the cause, of the phenomenon.[2]

D. S. Robinson[3] calls Mill's method of difference the method of experiment *par excellence.*

Statistical experiments in education in the past have largely been carried on by the method of experimental-control groups. While much valuable information has been obtained by this means, the method is open to some grave weaknesses. The chief one of these is the difficulty of keeping the experimental group on a comparable basis with the control group. The cause of the difficulty is that the experimental group is subjected to a more or less artificial treatment in subject-matter method, teacher personnel, or a combination of these.

The present investigation, then, in using the scientific method of difference, has attempted a somewhat new departure in educational experimentation.

[1] Mill, John Stuart. *System of Logic*, Vol. I, Book III. Harper and Brothers, New York, 1850 ed.   [2] *Ibid.*, Chap. VIII, p. 225.
[3] Robinson, D. S. *The Principles of Reasoning*, p. 264. D. Appleton and Company, New York, 1924.

In nearly all performance tests in school subjects there are a number of abilities that enter into each test. These abilities depend upon such factors as intelligence, previous practice in the process (immediate or remote), the emotional status of the subject, his physical condition, his reaction time—all of which are difficult to control. In any natural situation, a given ability can be observed only as mixed with and conditioned, to a greater or less extent, by one or more of these factors. Before it can be measured, it must be separated from them. In the present investigation, the variables to be measured are the skills required in the subtraction techniques, and the problem is to isolate these abilities.

CONSTRUCTION OF THE TESTS

To effect a solution to this problem of isolating the variable, first two tests were constructed so that the only ability called into use in the second test over the abilities used in the first test was that required by the subtraction technique.

An illustration will make this clear. If subtraction in the example at the left is done by the decomposition method, the $\frac{6234}{3786}$ procedure is 6 from 14, 8 from 12, 7 from 11, and 3 from 5; that is, the subject performs these subtractions: $\frac{14}{6}$ $\frac{12}{8}$ $\frac{11}{7}$ and $\frac{5}{3}$ and, in addition to this, he performs the extra operation of borrowing (so-called), necessary to this method. The extra time it takes for subtracting $\frac{6234}{3786}$ over the time it takes for the simple facts $\frac{14}{6}$ $\frac{12}{8}$ $\frac{11}{7}$ and $\frac{5}{3}$ is clearly due to the extra operation or technique of borrowing.

In the same example, working by the equal additions or the Austrian method, the subject performs the following simple subtractions: $\frac{14}{6}$ $\frac{13}{9}$ $\frac{12}{8}$ and $\frac{6}{4}$ and, in addition to these, he performs the extra operation required in these methods. In these methods also, the time required for the example over and above the time required for the separate facts of $\frac{14}{6}$ $\frac{13}{9}$ $\frac{12}{8}$ and $\frac{6}{4}$ is due to the extra operation which these techniques demand.

The two tests were thus constructed (see Tests 1D and 2D), the first containing all of the 100 separate subtraction facts and the second these same 100 facts in the same order but put together in examples of subtraction so as to require the one extra operation demanded by the method used.

THE TESTS

The 100 Subtraction Facts (Test 1D)

9	5	11	7	11	13	12	5	7	8
1	3	5	1	9	8	5	1	7	0
12	6	16	8	14	1	10	9	9	7
9	3	7	3	5	1	2	5	4	4
11	15	11	8	5	8	3	14	6	12
7	9	8	7	0	6	1	7	2	6
12	17	7	2	3	13	4	17	6	13
7	9	2	1	0	4	1	8	0	5
2	10	9	9	6	12	13	14	0	1
2	8	8	7	4	4	9	9	0	0
8	6	16	8	13	11	13	5	4	2
1	5	9	4	6	6	7	4	2	0
14	7	11	4	11	3	10	9	9	8
6	5	4	0	2	3	1	3	6	5
10	18	10	5	4	9	3	16	7	15
4	9	6	5	4	2	2	8	3	8
10	15	6	7	7	12	5	14	8	15
5	7	1	6	0	3	2	8	2	6
6	10	9	9	8	11	12	10	4	10
6	3	9	0	8	3	8	7	3	9

Subtraction Examples (Test 2D)

5 9	8 1	1 4 1	7 6 2	7 2 8	2 4 9 6	8 0 0 0	5 9 2 6 1
3 1	1 5	8 9	7 1 5	3 9 0	1 5 3 7	4 4 5 2	7 8 9 7

3 8	7 4	1 3 2	2 8 7	5 3 3	3 3 7 7	7 0 0 0	1 1 5 4 2
1 6	2 7	7 6	1 2 9	1 4 0	2 5 0 8	4 7 8 8	9 9 4

6 8	9 6	1 2 3	4 6 3	8 4 2	4 1 5 1	9 0 0 0	4 6 1 9 0
5 1	4 9	6 6	2 4 7	5 6 0	3 2 0 4	5 6 3 1	4 5 6 9 4

3 9	8 6	1 1 5	7 7 5	6 2 7	7 5 9 4	9 0 0 0	1 0 5 1 3 1
2 2	3 8	5 8	6 1 7	2 3 0	6 6 2 8	8 0 9 3	9 3 7 8 3

The 100 Subtraction Facts (Test 1E)

7	8	12	8	15	12	11	1	3	9
3	2	3	7	7	4	7	0	1	0

16	6	15	9	14	4	15	0	10	9
8	5	6	3	8	3	8	0	1	9

11	7	7	6	5	9	3	14	5	18
9	0	7	3	4	5	3	6	3	9

11	13	5	8	6	16	6	11	7	11
3	7	5	4	0	9	6	2	6	4

9	10	10	10	9	11	3	4	8	8
2	3	4	7	6	6	0	1	6	0

2	9	13	3	13	13	14	2	6	5
2	4	4	2	9	5	9	1	2	0

14	9	12	9	17	4	17	10	10	7
7	1	5	7	8	2	9	9	8	5

12	12	4	5	7	8	1	12	8	15
7	6	4	2	2	3	1	9	8	9

The 100 Subtraction Facts (Test 1E)—Continued

12	11	6	7	4	11	9	14	8	16
8	8	1	1	0	5	8	5	1	7

6	10	10	10	8	13	2	13	7	5
4	2	6	5	5	8	0	6	4	1

Subtraction Examples (Test 2E)

8 7	8 2	1 2 5	3 1 1	6 6 9	4 4 9 5	9 0 0 5	5 6 7 7 1
2 3	6 3	3 7	9 7	4 8 0	2 8 2 6	8 0 9 8	4 3 6 9 9

3 9	5 4	1 1 8	8 5 3	6 6 6	9 1 7 1	9 0 0 0	8 8 4 3 1
3 5	2 6	2 9	4 4 7	5 9 0	1 4 5 2	5 6 3 3	6 0 9 6

9 2	3 3	1 3 3	6 2 4	9 4 5	4 7 9 2	7 0 0 7	7 5 4 2 2
4 2	1 4	4 9	2 0 9	7 0	1 8 6 5	4 7 8 9	2 2 3 5 7

1 8	8 2	1 2 5	7 6 1	9 1 4	6 6 8 4	8 0 0 0	5 7 3 2 3
1 3	7 9	7 9	1 0 8	7 5 0	3 7 0 5	4 4 5 2	1 3 5 9 8

When the time taken for Test 1D, which is made up of the separate 100 subtraction facts, is subtracted from the time taken for Test 2D, which contains the same subtraction facts but requires the extra operation of the subtraction technique, the effects produced by such factors as physical, emotional, and mental condition are eliminated by reason of being common to the results of both tests, 1D and 2D. The difference between the number of errors made on the second test and the number made on the first test is likewise uninfluenced by the personal equation factors. This procedure is known as the method of differential testing.

A criticism may be offered here in that the subject using the D technique would not use the same simple facts as the subject using the E technique or the A technique, when doing the same subtraction example. This is illustrated in the example just given, $\frac{6234}{3786}$. The D subject says, 6 from 14, 8 from 12, 7 from 11, 3 from 5; whereas the E subject says, 6 from 14, 9 from 13, 8 from 12, 4 from 6; and the A subject says, 6 and ? are 14, 9 and ?

are 13, 8 and ? are 12, 4 and ? are 6? This objection is answered by the fact that, since both method groups use all of the 100 subtraction facts in Test 2 as well as in Test 1, they cover the same ground as each of the other method groups. Another criticism that has been offered is not only that the same facts should be covered in the first test as in the second, but that the order of the facts should be the same in the two tests. This could evidently not be done if the three groups used the same two tests. To offset this criticism, two sets of tests had to be made. Besides the set called 1D and 2D, another set called 1E and 2E was devised (see pages 44 and 45). Tests 1D and 2D were to be used by the D group and Tests 1E and 2E by the E group and A group, the examples having been so constructed that the order of the facts was the same in Test 1D as in Test 2D, and in Test 1E as in Test 2E.

An anticipated difficulty in administering the tests appeared. They were made so that each set of tests should be used by its respective method group, but seldom if ever was there found a class all of whose members used the same method. The general rule was a mixture of two or three methods in the same room. To overcome this difficulty of administration, and also the probable criticism that the two groups were not taking the same tests, although the tests were equivalent, it was decided to give each group both sets of tests. This would obviate the difficulty mentioned above and at the same time enable the writer to obtain a measure of reliability from the two sets of tests by the same group. Furthermore, it was the special object of this experiment not to do any artificial sectioning of classes or carry on any procedure other than the ordinary procedures in any classroom, but simply to give the tests to all alike in their natural situation and in the same way. Thus all groups took the same tests, 1D and 2D and also 1E and 2E.

It should be added that Tests 2D and 2E, consisting of four rows of examples, were constructed so as to contain all of the types of subtracting difficulties, arranged from easy to difficult in each row, the difference in difficulty depending upon the number and kinds of steps involved. Each row has twenty-five subtractions in it.

ADMINISTRATION OF THE TESTS

The returns from questionnaires sent to 300 principals gave information regarding which method predominated in the respective schools. After checking methods used by the pupils, permission was obtained from eight principals to give tests in their schools. The writer gave all of the tests personally. The first day both Test 1D and Test 2D—following immediately—were given in the same class period. In most cases the second set of tests, 1E and 2E, was given to the same pupils the same week or the following week. In almost all of the cases both sets of tests were given within the same four weeks' period. The E tests were given first if the majority of the class used the E or A method. At some time between the administration of the tests each member of each class was asked to record on a separate sheet of paper just what he said (thought) when he subtracted these two examples, $\frac{82}{37}$ and $\frac{600}{146}$ (see pupil examples, one from each method, given below). The two different examples were asked for to find out whether the pupil was consistent in using the same method in both examples. Only one out of some 1,200 pupils used different methods on these two examples. This pupil, most likely, had been taught the D method and had picked up the E method on the easy examples. In this case, where the zeros occurred she had reverted to the more familiar method. As this pupil could not be classified, her results were not used; her record was very poor.

The individual pupils were thus classified into three groups on the basis of the method shown on these individual papers. For the purpose of getting mental age comparisons, an intelligence test (the McCall Multi-Mental Scale) was given to each pupil shortly after the two sets of subtraction tests had been given. Thus each pupil was subjected to five tests at three different sittings. In all, more than 6,000 pupil-tests were given. They were all given personally by the writer. This amounted, on the average, to testing one class each week for nearly two school years.

Pupil N. S.		*Pupil E. L.*		*Pupil F. S.*	
Method D		Method E		Method A	
82	7 from 12 = 5	82	7 from 12 = 5	82	7 and 5 = 12
−37	3 from 7 = 4	−37	4 from 8 = 4	−37	4 and 4 = 8
45		45		45	
600	6 from 10 = 4	600	6 from 10 = 4	600	6 and 4 = 10
−146	4 from 9 = 5	−146	5 from 10 = 5	−146	5 and 5 = 10
454	1 from 5 = 4	454	2 from 6 = 4	454	2 and 4 = 6

As indicated before in another connection, a fundamental principle in conducting this experiment was not to upset any class by changing teacher, period, or procedure, so that the tests could be obtained under ordinary conditions in each classroom. An arithmetic period was usually used. The writer came into the room and asked the pupils if they would like to play a little arithmetic game, the hardest fact of which was 9 from 17. Test 1D or 1E was then distributed face down and each pupil was instructed when the signal came to turn the paper and to do the examples as accurately as he could. The children were told to take as much time as they needed, not hurrying to get through, and to hand in their papers when they had worked all the examples.

After the signal to start had been given, the writer sat at the desk, noting time of beginning by his watch and marking the elapsed time in 10-second intervals on each paper as it was handed in. As soon as Test 1 was finished by all, Test 2 was distributed and administered in the same way. The two tests together required from 30 to 40 minutes of time, since time was given the slower pupils to finish the entire test.

Scoring the Tests

With the help of two graduate students, the writer scored all of the tests. Each subtraction fact missed on Test 1 was encircled. Likewise each fact missed as a part of an example in Test 2 was encircled, some examples yielding more than one error. By this procedure it was possible to determine whether a fact which was missed in Test 1 was also missed in Test 2. This yielded the information in regard to which errors made in Test 1 were accidental and which errors made in Test 2 were due to the subtraction technique.

A very surprising result was found in this connection. A careful check of errors in both tests from a number of pupils was made to see if the same error occurred in both. Only 15 of 155 errors examined from Test 1 were found also in Test 2 of the same students. In Test 2, taken by the same subjects, there were 716 errors. That is to say, a subtraction fact missed in Test 1 was more often correct than incorrect in Test 2, the ratio being about 10 to 1. This showed that errors in Test 1 were largely accidental. It showed furthermore that errors in Test 2 were due largely to the subtraction technique. To illustrate on a small scale from one example: Suppose a subject missed in Test 2 the two middle facts which are encircled in the example at left, but

in Test 1, out of the four simple facts, $\dfrac{16\ 8\ 14\ 1}{7\ 3\ 5\ 1}$ that go

to make up the example, he missed only $\dfrac{16}{7}$. This would

indicate that $\dfrac{16}{7}$ was an accidental error because it was not missed in the example of Test 2. It would show further that the errors on the middle two facts in the example were probably due to the subtraction technique. Again these errors proved that making the order of the facts in Test 1 the same as the order in Test 2 was an overzealous precaution that did neither good nor harm.

It will be admitted that some of the errors in Test 2 may be due to accident. This should not vitiate the results, however, because the accidental errors in Test 1 should balance the accidental errors in Test 2. Even if one is inclined to doubt this, he will certainly admit that the ratio of accidental errors to real errors in either Test 1 or Test 2 would not differ greatly within the results of method D, E, or A.

Of the two bases of comparison, time and accuracy, however, time is the more significant. While there were many accidental errors in accuracy, there could be no accidental errors in time. Hence the time increment score on Test 2 is a good index of the difficulty of the subtraction technique used.

V

THE DATA PRESENTED

RELIABILITY OF THE TESTS

SINCE most of the pupils took both Test 2D and Test 2E, the results gave opportunity for measuring the reliability of the tests. Each test, as has been stated before, contained all of the 100 subtraction facts, and each subtraction fact was presented but one time.

The correlations between results of Test 2D and results of Test 2E for both time and accuracy were computed. Four separate correlations were computed, one for the third and fourth grades, one for the fifth grades, one for the sixth and seventh grades, and one for the eighth grades. These are shown in Table I.

These rather high coefficients of correlation between the results of Tests 2D and 2E show that the abilities measured by the two tests were very closely related.

The results show higher reliability in time than in accuracy. This is perhaps due to the fact that some errors are accidental.

Furthermore, the coefficients are highly reliable, as is shown by the relative sizes of the coefficients and their probable errors.

TABLE I

CORRELATIONS BETWEEN RESULTS OF TEST 2D AND TEST 2E

Grade	N	Accuracy		Time	
		r	P. E.$_r$	r	P. E.$_r$
Grades 3 and 4	31	.66	±.08	.70	±.06
Grades 5	163	.58	±.04	.70	±.03
Grades 6 and 7	159	.62	±.03	.80	±.02
Grades 8	87	.58	±.05	.76	±.03

COMPARABILITY OF THE VARIOUS GROUPS

As the method employed in this experiment is that of differential testing, it would seem unnecessary to have comparable groups. However, to render comparisons between the results of Test 2[1] from the various groups absolutely reliable, all groups were made equivalent with respect to the following bases: mental ages, intelligence quotients, results in errors on Test 1, and results in time on Test 1.[2] This was effected by making the means and standard deviations in the three groups nearly the same. This called for a

TABLE II

MENTAL AGES AMONG THE VARIOUS GROUPS

M. A.*	Frequencies			
	D	E	A	E + A†
17 and above‡......	25	11	7	18
16.................	13	5	5	10
15.................	37	25	10	35
14.................	47	36	13	49
13.................	73	41	26	67
12.................	93	58	27	85
11.................	82	55	30	85
10.................	113	75	53	128
9.................	36	33	14	47
8.................	5	3	1	4
7.................	2			
N.................	526	342	186	528
M.................	12.0	11.8	11.8	11.8
S. D.	2.14	2.06	2.07	2.06

* M. A. scores are midpoints, 12 denoting the interval 11.5–12.5.

† The E + A Group is made up of the E and A groups combined and can be considered as a group that increases the subtrahend figure as contrasted with the D group that increases the minuend figure in subtraction.

‡ Taken as 17 in the computations.

[1] When results of Test 2 are mentioned it is understood to mean the average of the results from Test 2D and Test 2E, for each pupil took both Tests 2D and 2E.

[2] Results in errors and time in Test 1 likewise mean average results from Test 1D and Test 1E.

To avoid wordiness and ambiguity in the headings of the tables that follow, Test 1 means the average of Test 1D and Test 1E; likewise, Test 2 means the average of Test 2D and Test 2E.

delicate manipulation in discarding from the distributions that had the high means and large S.D.'s certain individual cases. Extreme care had to be exercised here, as three groups were under consideration, and whenever one card from one individual was thrown out it affected all four variables at the same time, viz., M.A.'s, I.Q.'s, errors in Test 1, and time in Test 1. Happily, in this instance, the D distribution had both the highest means and the largest S.D.'s in most of the variables so that the adjustment was made without an undue amount of labor.

The result of this balancing on the four bases mentioned is shown in Tables II, III, IV, and V.

Table II shows that the groups are highly comparable on the basis of mental age, as revealed by the respective mental age means of 12.0, 11.8, 11.8, and 11.8 years. It also shows that the character of the distributions about the means is very much the same in each of the groups, as shown by standard deviations of 2.14, 2.06, 2.07, and 2.06 respectively.

TABLE III

INTELLIGENCE QUOTIENTS AMONG THE VARIOUS GROUPS
(I. Q.'s FROM McCALL'S MULTI-MENTAL SCALE)

I. Q.*	Frequencies			
	D	E	A	E + A
165–174............	4		1	1
155–164............	4	1		1
145–154............	7	4	5	9
135–144............	14	10	4	14
125–134............	43	21	14	35
115–124............	67	50	25	75
105–114............	104	79	27	· 106
95–104............	117	77	46	123
85– 94............	97	51	40	91
75– 84............	52	37	18	55
65– 74............	15	10	6	16
55– 64............	2	2		2
N.................	526	342	186	528
M.................	104.9	104.2	103.9	104.1
S. D..............	18.7	17.4	18.4	17.7

* I. Q. scores are integral limits. 95–104 indicates 95–104.9.

Table III shows that the groups are highly comparable on the basis of intelligence quotients, as revealed by the respective I.Q. means of 104.9, 104.2, 103.9, and 104.1. It also shows that the character of the distributions about the means is very much the same in each of the groups, as shown by standard deviations of 18.7, 17.4, 18.4, and 17.7 respectively.

It will be noted that the small difference in means among the groups is in favor of the D group. This is true with respect to mental ages as well as I.Q.'s.

Table IV shows that the groups are highly comparable on the basis of errors made in Test 1, as revealed by the respective error means of 1.31, 1.28, 1.36, and 1.31. It also shows that the character of the distributions about the means is very much the same in each of the groups, as shown by standard deviations of 1.41, 1.44, 1.44, and 1.44 respectively.

TABLE IV

NUMBER OF ERRORS MADE IN TEST 1 BY THE VARIOUS GROUPS

Number of Errors*	Frequencies			
	D	E	A	E + A
11...............			1	1
10...............	3	1		1
9...............				
8...............	1	3		3
7...............	1	1		1
6...............	4	5	2	7
5...............	6	3	2	5
4...............	17	9	6	15
3...............	22	7	9	16
2...............	54	40	15	55
1...............	123	77	46	123
0...............	338	239	105	344
N...............	569†	385†	186	571
M...............	1.31	1.28	1.36	1.31
S. D.............	1.41	1.44	1.44	1.44

* Scores are low points of interval. Score of 5 indicates 5.0–5.9.

† Tables IV and V include 43 adults who were not included in Tables II and III since no M. A. or I. Q. was measured for them. Hence the totals 569 in group D and 385 in group E are each 43 greater than the corresponding totals in Tables II and III.

It is of interest also to note the large number of perfect papers, which reduces the mean error to the low figure of 1.3. Besides having the same standard deviation, 1.4, the distributions are likewise similarly skewed, as shown by the small variation between frequency numbers in the various score groups in the D and E + A distributions. This adds to the reliability of results from later tables where scores from Test 2 are compared.

Table V shows that the groups are highly comparable on the basis of time taken on Test 1, as revealed by the respective time means of 4.82, 4.81, 4.79, and 4.80 minutes. It also shows that the character of the distributions about the means is very much the same in each of the groups, as shown by standard deviations of 2.02, 2.13, 1.67, and 2.00 respectively.

(A fact incidentally revealed by Table V is that a pupil with an M.A. of 12 years, which is normally a sixth grade pupil, and an

TABLE V

TIME TAKEN ON TEST 1 BY THE VARIOUS GROUPS

Number of Minutes*	Frequencies			
	D	E	A	E + A
14................	2			
13................	1	1		1
12................	1			
11................	1	3		3
10................	8	5	1	6
9................	9	9	3	12
8................	22	12	5	17
7................	36	32	12	44
6................	53	36	16	52
5................	92	62	37	99
4................	106	51	40	91
3................	142	98	54	152
2................	83	63	16	79
1................	13	13	2	15
N................	569	385	186	571
M................	4.82	4.81	4.79	4.80
S. D..............	2.02	2.13	1.67	2.00

* Scores are low points of intervals.

average I.Q. takes about 5 minutes to give the answers to the 100 subtraction facts in written form. This is longer than many authors of textbooks expect.)

Table VI is a summary of Tables II, III, IV, and V. It brings the means and standard deviations in the four distributions into closer proximity for easy comparison.

TABLE VI

SUMMARY: EQUIVALENCE OF THE GROUPS

Group	N	M	S. D.
Mental Ages			
D..........................	526	12.0	2.14
E..........................	342	11.8	2.06
A..........................	186	11.8	2.07
E + A....................	528	11.8	2.06
Intelligence Quotients			
D..........................	526	104.9	18.7
E..........................	342	104.2	17.4
A..........................	186	103.9	18.4
E + A....................	528	104.1	17.7
Errors in Test 1			
D..........................	569	1.31	1.41
E..........................	385	1.28	1.44
A..........................	186	1.36	1.44
E + A....................	571	1.31	1.44
Time in Minutes on Test 1			
D..........................	569	4.82	2.02
E..........................	385	4.81	.2.13
A..........................	186	4.79	1.67
E + A....................	571	4.80	2.00

DIFFERENCES IN THE SUBTRACTION RESULTS COMPARED

The equivalence of the groups has thus been established on the bases of mentality and original ability in the 100 basic subtraction facts.

If the number of errors made and the time taken on Test 1 are subtracted from the number of errors made and the time taken on Test 2 by each group, the difference found is due to the subtraction technique used by that group. The smaller that difference is, the better the technique used.

These differences are reported in Tables VII and VIII.

TABLE VII

INCREASE IN ERRORS IN TEST 2 OVER TEST 1

Increase in Errors	Frequencies			
	D	E	A	E + A
29–30	2	2		2
27–28	3	1		1
25–26	3		1	1
23–24	3	1		1
21–22	8	3	3	6
19–20	7	8	1	9
17–18	12	5	1	6
15–16	16	5	3	8
13–14	17	4	5	9
11–12	21	6	7	13
9–10	35	19	5	24
7– 8	45	28	9	37
5– 6	55	54	29	83
3– 4	99	69	49	118
1– 2	136	111	47	158
−1– 0	97	66	25	91
−3– −2	10	3	1	4
N	569	385	186	571
M	5.73	4.84	4.92	4.87
S. D.	6.04	5.20	4.74	5.06
P.E.$_M$	.171	.178	.234	.143

Table VII shows that each member of the D group made an average of 5.73 more errors in Test 2 than in Test 1; whereas each member of group E made on an average only 4.84 more errors in Test 2 than in Test 1. Likewise each member of group A made only 4.92 more errors in Test 2 than in Test 1 and each member of groups E and A together made 4.87 more errors in Test 2 than in Test 1.

What conclusion can be drawn from these differences? In each group there are more errors in Test 2 than in Test 1, but this increase in errors is not the same among the groups, as noted from Table VII. The difference in the increases then must be due to the differences in the methods used. Furthermore, the group that made the greatest number of additional errors in Test 2 over Test 1 used the least efficient technique of subtraction as far as accuracy is concerned.

If that assumption is correct—and there seems to be no reason to believe it is not—then the D method of subtraction, by reason of its own intrinsic nature, produces more errors than either of the other two methods. Also, the equal additions method and the Austrian method are about equal in efficiency as regards accuracy, with an insignificant margin in favor of the equal additions method.

TABLE VIII

INCREASE IN TIME ON TEST 2 OVER TEST 1

Increase in Minutes*	Frequencies			
	D	E	A	E + A
13..............	4			
12..............	2	2		2
11..............	2			
10..............	4	2		2
9..............	4	4		4
8..............	12	4	1	5
7..............	16	13	1	14
6..............	17	14		14
5..............	35	18	4	22
4..............	55	29	8	37
3..............	76	40	18	58
2..............	138	72	42	114
1..............	152	120	60	180
0..............	50	59	50	109
−1..............	2	8	2	10
N..............	569	385	186	571
M..............	3.19	2.76	1.91	2.48
S. D.	2.33	2.24	1.39	2.04
P.E.$_M$	.066	.077	.068	.058

* Scores are low points of intervals as before.

Table VIII shows that the extra time taken for Test 2 over Test 1 by the D group was 3.19 minutes; by the E group it was only 2.76 minutes; by the A group it was only 1.91 minutes; and by the E and A group it was 2.48 minutes.

The same line of reasoning holds here as that following table VII, the difference being that Table VIII concerns time, whereas Table VII deals with errors. From the above figures, it is readily seen the D method is the least efficient and the A method the most efficient on the basis of time.

Table IX is a summary of Tables VII and VIII, and draws into closer relation the means, standard deviations, and probable errors of the means of those tables. In Table X these differences are treated statistically.

TABLE IX

SUMMARY: MEAN INCREASES IN ERRORS AND IN TIME
ON TEST 2 OVER TEST 1

Group	Number	Mean	S. D.	P. E.$_M$
In Errors				
D.	569	5.73	6.04	.171
E.	385	4.84	5.20	.178
A.	186	4.92	4.74	.234
E + A.	571	4.87	5.06	.143
In Time				
D.	569	3.19	2.33	.066
E.	385	2.76	2.24	.077
A.	186	1.91	1.39	.068
E + A.	571	2.48	2.04	.057

Table X sets forth the following facts:

1. Method D causes 18.4% more errors than method E and at the same time requires 15.5% more time.

2. Method D causes 16.4% more errors than method A and at the same time requires 67% more time.

3. Method D causes 17.7% more errors than methods E and A combined and at the same time requires 28.6% more time.

4. Method E causes 1.6% less errors than method A but requires 44.5% longer time.

TABLE X

DIFFERENCES IN THE MEAN INCREASES IN ERRORS AND THE MEAN
INCREASES IN TIME ON TEST 2 OVER TEST 1, EVALUATED IN TERMS OF
PERCENTAGES AND CRITICAL RATIOS

Groups	$M_1 - M_2$	P. E. diff.	% diff.*	C. R.†
In Errors				
D *vs.* E............	.89	.246	18.4	3.61
D *vs.* A............	.81	.290	16.4	2.79
E *vs.* A............	−.08	.294	1.63	.271
D *vs.* E + A........	.86	.223	17.7	3.86
In Time				
D *vs.* E............	.43	.101	15.5	4.25
D *vs.* A............	1.28	.094	67.0	13.6
E *vs.* A............	.85	.102	44.5	8.3
D *vs.* E + A........	.71	.088	28.6	8.1

* Per cent by which the increase in number of errors made by the first group was larger than the increase in number of errors made by the second group.
† The critical ratio is the ratio of a difference to its probable error. A critical ratio of 3 or more indicates statistical significance.

Table X, furthermore, shows that the above comparisons are all statistically significant except the comparison of errors between methods E and A. This means that the samplings of populations to which the tests were given were such that should the experiment be repeated with other population samplings it would show results in the same direction favoring the E and A methods.

VI

THE SECONDARY EXPERIMENT

COMPARISON BETWEEN GROUP D AND GROUP E + A IN TIME ONLY, ERRORS BEING CONSTANT

THE foregoing data have yielded comparisons both in errors and in time. They would give added information if comparison could be made in terms of one variable only. This was done by selecting only those cases from each group who made no errors in Test 1 and only a limited number of errors in Test 2. In this way the most competent as well as the most careful members of each group were compared. It is natural that most individuals want to be accurate in their subtraction, and in trying to be accurate a certain amount of time is required. Thus, since other things are equal, if the subtraction techniques which are being studied in this experiment are intrinsically different in difficulty, that fact ought to be manifested in the different amounts of time required for Test 2 by the groups using the various techniques.

Comparisons were made first between two groups that made no errors in Test 1 and no errors in Test 2; then between two groups that made no errors in Test 1 and from one to four errors in Test 2.

The two main groups only were compared, that is, the D group *vs.* the E + A group, as the number of cases was greatly reduced.

(In examining these groups for comparability it was found that they were already equivalent in mental ages and intelligence quotients. In making them equivalent on the basis of time on Test 1 but three cases were discarded, two from the D group and one from the E + A group.)

Table XI is only a summary comparing the means and S.D.'s in the two groups now made equivalent on the basis of M.A.'s, I.Q.'s, errors in Test 1, errors in Test 2, and time on Test 1.

If the pairs of numbers in each column above are compared, the reader will note the nearly equal size of means and S.D.'s in both comparisons on the five bases indicated.

TABLE XI

EQUIVALENCE OF THE GROUPS

	First Comparison*			Second Comparison†		
	N	M	S. D.	N	M	S. D.
In M. A.'s						
Group D.........	41	13.3	2.14	144	12.4	2.29
Group E + A.....	39	13.0	2.24	144	12.1	2.03
In I. Q.'s						
Group D.........	41	104.8	10.2	144	108.5	1.14
Group E + A.....	39	105.0	18.9	144	108.5	1.14
In Errors in Test 1						
Group D.........	52‡	0.0	0.0	156‡	0.0	0.0
Group E + A.....	52	0.0	0.0	158	0.0	0.0
In Errors in Test 2						
Group D.........	52	0.0	0.0	156	2.72	1.07
Group E + A.....	52	0.0	0.0	158	2.75	1.01
In Time on Test 1						
Group D.........	52	3.64	1.22	156	4.21	1.64
Group E + A.....	52	3.56	1.21	158	4.23	1.62

* Between groups who had no errors in either Test 1 or Test 2.

† Between groups who had no errors in Test 1 and from 1 to 4 errors in Test 2.

‡ The larger number here is due to the inclusion of adults who were not included in the M. A. and I. Q. measurements.

These newly equated groups will now be compared in time on Test 2 only. Tables XII, XIII, and XIV show the results.

Table XII reveals an interesting result and one that should be highly significant. Here are two groups which are very accurate or very careful or both, as shown by the fact that neither made any errors in either the 100 basic subtraction facts in Test 1 or the 25 examples which involved the subtraction technique. One group used the decomposition method, which decreases the minuend figure by one; the other group employed the equal additions method or the Austrian method, both of which increase the subtrahend figure by one. Each group took about the same length of time on the 100 basic subtraction facts. These same 100 subtraction facts were in-

TABLE XII

INCREASE IN TIME ON TEST 2 OVER TIME ON TEST 1

(By the groups that had no errors in Test 1 or Test 2, and that were equated in time on Test 1 and also in M. A.'s and I. Q.'s)

Increase in Minutes	Frequencies	
	D	E + A
4...	5	1
3...	6	4
2...	14	7
1...	19	18
0...	8	21
−1...		1
N...	52	52
M...	2.14	1.41
S. D.	1.16	1.04
P.E.	.109	.0974

cluded in Test 2, from which comparisons are now made. These two groups were also of equal mentality.

Yet the two groups differ in the amount of time required for Test 2.

If the assumption is correct that the extra time required for Test 2 over Test 1 is due to the subtraction technique, then there is a direct causal connection between the method of subtraction and the extra time consumed on Test 2.

The group using method D consumed 2.14 minutes of extra time; whereas the group using methods E and A used but 1.41 minutes of extra time.

This seems to indicate that the decomposition method is the least efficient method.

To corroborate or to put in doubt the conclusion drawn from Table XII, a second comparison was made with a larger number in each group. This time the groups included all those from each method who made no errors in Test 1 and but one to four errors in Test 2. These were also accurate and careful groups. The results in time from Test 2 by these groups are shown in Table XIII.

TABLE XIII

INCREASE IN TIME ON TEST 2 OVER TIME ON TEST 1

(By the groups that had no errors in Test 1, and 1 to 4 errors in Test 2, and that were equated in time in Test 1 and also in M. A.'s and I. Q.'s)

Increase in Minutes	Frequencies	
	D	E + A
13................................	1	
12................................	1	
11................................	1	
10................................		
9................................	1	2
8................................	2	1
7................................	4	1
6................................	5	2
5................................	5	1
4................................	12	9
3................................	23	18
2................................	40	28
1................................	46	52
0................................	15	40
−1................................		3
−2................................		1
N................................	156	158
M................................	2.98	2.04
S. D.	2.25	1.71
P. E.	.122	.091

The same reasoning can be employed with regard to Table XIII as was used in the discussion of Table XII.

The extra time required on Test 2 by group D is 2.98 minutes and the extra time required by group E + A is only 2.04 minutes. Almost 50% more time was taken by group D than by group E + A.

Conclusions for Table XIII are the same as for Table XII.

Table XIV displays the data for a combination of the groups represented in Tables XII and XIII. It shows that the D group takes 2.77 minutes more to do Test 2 than Test 1, whereas the E + A group requires only 1.88 minutes more to do Test 2 than Test 1.

As time is a better index than errors (some errors being acci-

TABLE XIV

INCREASE IN TIME ON TEST 2 OVER TIME ON TEST 1

(By the groups that had no errors in Test 1 and 0 to 4 errors in Test 2 and that were equated in time on Test 1 and also in M. A.'s and I. Q.'s)

Increase in Minutes	Frequencies	
	D	E + A
13.................................	1	
12.................................	1	
11.................................	1	
10.................................		
9.................................	1	2
8.................................	2	1
7.................................	4	1
6.................................	5	2
5.................................	5	1
4.................................	17	10
3.................................	29	22
2.................................	54	35
1.................................	65	70
0.................................	23	61
−1.................................		4
−2.................................		1
N.................................	208	210
M.................................	2.77	1.88
S. D.	2.06	1.62
P. E.	.096	.075

dental), the results of Tables XII, XIII, and XIV cannot be passed over lightly in this experiment. There seems to be no doubt that the decomposition method is the least efficient method of subtraction.

Table XV summarizes the results of Tables XII, XIII, and XIV to facilitate comparisons between the time means of groups that were equivalent in other respects. Table XVI will treat these differences statistically for final conclusions.

Table XVI presents the following comparison:

Of those that had no errors in either Test 1 or Test 2 the D group required 51.7% more time than did the E + A group.

TABLE XV

SUMMARY: MEAN INCREASES IN TIME ON TEST 2 OVER TIME ON TEST 1
(OTHER VARIABLES BEING CONSTANT)

	N	M	S. D.	P. E.$_M$
*In First Comparison**				
Group D..........	52	2.14	1.16	.109
Group E + A......	52	1.41	1.04	.0974
In Second Comparison†				
Group D..........	156	2.98	2.25	.122
Group E + A......	158	2.04	1.71	.091
In Both Comparisons Combined				
Group D..........	208	2.77	2.06	.096
Group E + A......	210	1.88	1.62	.075

* Between groups who had no errors in Test 1 or Test 2.
† Between groups who had no errors in Test 1 and 1 to 4 errors in Test 2.

TABLE XVI

DIFFERENCES IN THE MEAN INCREASES IN TIME ONLY ON TEST 2 OVER
TEST 1, EVALUATED IN TERMS OF PERCENTAGES AND CRITICAL RATIOS

	$M_D - M_{E+A}$	P. E. diff.	% diff.	C. R.
In first comparison*.......	.73	.143	51.7	5.1
In second comparison†.....	.94	.151	46.2	6.2
In combined comparisons‡..	.89	.122	47.3	7.3

* Between the groups who made no errors in Test 1 and no errors in Test 2.
† Between the groups who made no errors in Test 1 and 1 to 4 errors in Test 2.
‡ Between the groups who made no errors in Test 1 and 4 or fewer errors in Test 2.

Of those that had no errors in Test 1 and from 1 to 4 errors in Test 2, the D group required 46.2% more time than the E + A group.

Of those that had no errors in Test 1 and from 0 to 4 errors in Test 2, the D group required 47.3% more time than did the E + A group.

Furthermore, all of these differences are statistically significant, the critical ratios being 5.1, 6.2, and 7.3 respectively.

VII

CONCLUSIONS AND PEDAGOGICAL IMPLICATIONS

Conclusions

THE conclusions with respect to this experiment, as drawn from data on the preceding pages, may be summarized as follows:

1. Other things being equal, the decomposition technique in subtraction of whole numbers is, by its own intrinsic nature, by far the poorest method to employ from the standpoint of both accuracy and time.

When compared with the equal additions method, the decomposition method produces 18% more errors and requires 15% more time.

When compared with the Austrian method, the decomposition method produces 16% more errors and takes 67% more time.

When compared with the equal additions method and the Austrian method combined, the decomposition method produces 17% more errors and requires 29% more time.

2. When the equal additions technique alone is compared with the Austrian technique there is but a slight difference in accuracy, 1.6%, in favor of the equal additions method. The difference is not statistically significant. The difference in time—44% in favor of the Austrian method—is large and is statistically significant. These two methods, then, are, so far as this experiment shows, about equal in accuracy, but the Austrian method is more rapid than the equal additions method.

3. When numbers of errors were constant, so that comparison could be made with respect to time only, the group using the decomposition method required 47% more time than did the combined group using the equal additions and Austrian methods.

4. Furthermore, the statistical significance of all of the above differences (except the one mentioned between equal additions and Austrian in accuracy) is such that should the experiment be re-

peated with different population samplings we could be practically certain that the results would be in the same direction, favoring the equal additions and Austrian methods over the decomposition method both in accuracy and in time.

Reinforcement of the above conclusions is offered by the fact that in the seven previously reviewed experiments[1] comparing the methods of subtraction the results were all in favor of the equal additions method or the Austrian method as against the decomposition method.

Pedagogical Implications

It has been shown that the equal additions method and the Austrian method are more accurate and more rapid than the decomposition method. What will an examination of the different methods offer, first, as to the causes that make one method better than another and, second, as to differences in facility in learning them?

First, when subtracting by the decomposition method in the example at the left, the pupil's thinking is as follows: "9 from 17 = 8"; but while he is doing the subtracting he must remember that the 4 has become 3. If he has difficulty in remembering that 9 from 17 is 8, he may forget that the 4 has become 3. He then thinks 7 from 13 (provided he remembers the 4 has become 3), but while he is doing that he has to remember that the 0 has become 9, and so on. In this method there is an element of memory with which the mind is encumbered while it is performing the subtraction.

    8047
    6679

When subtracting the above example by the Austrian method, no memory element is involved as the pupil now thinks "9 and 8 are 17." Then, and not until then, does he add the 1 to the 7, and as soon as he has added the 1 to the 7, making 8, he uses the 8 in thinking "8 and 6 are 14." His whole attention can be put upon the subtraction combination in question without having it divided by remembering something to be done to the next upper figure.

In using the equal additions method, the upper figures, as in the Austrian method, do not change. In the foregoing example the 7, 4, 0, and 8 of the minuend suggest 17, 14, 10, and 8, whereas in the decomposition method the 7 suggests 17, the 4 is changed to 13, the 0 is changed to 9, and the 8 is changed to 7. It is true that in the

---

[1] See last paragraph of Chapter III, page 40.

equal additions method the subtrahend figure is changed, but it is always by adding 1 more and nothing else. The change in the minuend figure under the decomposition method is more involved than the change in the subtrahend figure under the equal additions method, as is illustrated in the example on page 67. When, in the decomposition method, the pupil thinks "7 from 13" in the example just mentioned, the 4 has had a double change; first from 4 to 3, then from 3 to 13. That is to say, seeing 4 and thinking 13 is, to the pupil, a greater requirement than the corresponding change in the equal additions method, viz., seeing 7 and thinking 8.

Second, in an example like the one at the left, the decomposition method requires an additional skill not required by the equal additions method or the Austrian method, that of borrowing when
4000
there are two or more successive zeros in the minuend. This
2795
skill, as every teacher of arithmetic knows, must be specifically taught under the decomposition method. That is, a pupil may be able to do an example like $\frac{812}{687}$ and not be able to proceed in an example like $\frac{900}{687}$. He may be able to take the first step, 7 from 10, but he will almost certainly flounder on the second step. On the other hand, the pupil who uses either the equal additions method or the Austrian method sees no new skill in the second example above; in this example he merely thinks of each 0 as suggesting a 10.

Third, it is no longer doubted that addition is easier and performed more rapidly than subtraction. When we compare the three methods, we find upon close analysis each of the following steps as illustrated below:

81   Decomposition: 1 from 8, 1 + 10, 5 from 11, 6 from 7.
65   Equal additions: 1 + 10, 5 from 11, 6 + 1, 7 from 8.
     Austrian: 1 suggests 11, 5 and ? = 11, 6 + 1, 7 and ? = 8.

In comparing the three methods as applied in the example just given, we see that each method has four steps. In the decomposition method, one step is addition and three steps are subtraction. In the equal additions method, two steps are addition and two are subtraction. In the Austrian method, all four steps may be considered as addition. Hence, with respect to the number of additive steps, the Austrian method ranks first, the equal additions method ranks second, and the decomposition method ranks third.

There is one unexplored point, however, in the Austrian method. We do not yet know positively whether addition with one addend missing is as accurate and as rapid as addition with the sum missing. On the other hand, it would seem that additive subtraction would strengthen ordinary addition and the two would then be mutually reinforcing.

Now what have the three preceding comparisons to offer for the teaching of subtraction? From the first comparison, given on page 67, it is evident that either the Austrian method or the equal additions method requires less concentration and strain than the decomposition method since the mind is relieved from having its attention divided between a subtraction and a memory of something to be done to the next figure of the minuend. The second comparison, given on page 68, shows an actual saving of time in favor of the Austrian and equal additions methods since these methods avoid the necessity of teaching a major skill which is peculiar to the decomposition method, namely the skill of changing 4000 to 3000 + 900 + 90 + 10 when 4000 is the minuend in an example like 4000 − 2795. The third comparison shows greater facility in initial teaching, in favor of the Austrian method, since the 100 subtraction facts need not be taught as new facts but as slight modifications of addition facts already known.

Let us next examine the three methods from the standpoint of rationalization. If we are to rationalize fully when we teach subtraction, then the Austrian method has all the advantages, for all of the rationalization in this method consists in knowing that addition is the reverse of subtraction. The decomposition method comes next in ease of rationalization, and the equal additions method last. As a matter of fact, however, the pupil in the third grade, where subtraction is generally taught, does not, in the majority of cases, understand the rationalization that is given in connection with the teaching of subtraction. It may be beyond his capacity, or the difficulty may be in the way it is taught. At any rate, in most cases he learns the skill mechanically and performs it automatically. Without going into the psychological controversy of whether a pupil should be conscious of the reasons why he is doing what he is doing when he subtracts or whether he should merely do it as an automatic habit, if one method is found more efficient in use than another method, although the second method permits of more ease in rationalization, it is a question whether efficiency in later performance should be sacrificed for the sake of an initial stage of more facile explanation. Furthermore, one explains a process a few times only, whereas one computes many hundreds of times in a lifetime. This question of rationalization has been a cause of argument for many years among teachers of the two methods, decomposition and equal additions. Advocates of the decomposition method hold that their method is more easily explained, while advocates of the equal additions plan point out that experimental evidence of greater efficiency is all on their side. There has been no scientific proof that the decomposition method is more easily explained. This judgment is merely the opinion of teachers who have been taught that method and hence naturally understand that method better and consequently can explain it better. In all probability the equal additions method can be explained equally well, but most teachers have not yet learned how best to explain it.

Concerning rationalization, there can be no doubt that on first presentation of a new process, if students are mature enough to profit by rationalization, other things being equal, a rational procedure is far better than a mechanical one. In this respect the Austrian method is unique because its rationalization is tied up with its explanation. That is, in explaining the steps in the Austrian process of subtraction one unavoidably rationalizes the method because the steps in it are the steps of addition with one addend missing.

As there is no scientific evidence in regard to which of the two methods, the decomposition or the equal additions, can most easily be rationalized for young pupils, let it be granted for the sake of the argument that the decomposition method can be more easily rationalized than the equal additions method. The question then resolves itself into this: Shall a method that has been proved to be least efficient after it is learned continue to be taught just because it is more easily explained upon first presentation? Let the reader be the judge.

The relative merits of the Austrian and equal additions methods of subtraction as compared with the decomposition method have been ably summarized by Professor John C. Stone as follows:

For about thirty years many of our leading educators have seen the advantages of the additive method (Austrian). By a study of the psychology of learning they have seen that the skills of addition may be transferred to subtraction with but small loss. The ease on the attention must bring about great speed and accuracy, and thus effect economy in time both in the learning and in the use of subtraction. David Eugene Smith in *The Teaching of Elementary Mathematics*, 1900, said in discussing the three methods of subtraction in use, "But the third (the additive method) has the great advantage of using only the addition table in both addition and subtraction, and of saving much time in the operation."

While I have not kept complete data or published any study of the three methods in use, for over twenty years I have carried on careful experiments with children to determine the best method. I have also tested students entering the Normal School to try to find the relative efficiency in the use of the methods by adults.

I have found that children get the additive method much more quickly and use it more rapidly and accurately than they do the decomposition method. Also I find about the same difference among adults as those found by Johnson.[1] The difference varies with different classes and is often due to general native ability. But I find the equal additions method about 30% or 40% more accurate and rapid than the decomposition method, and the additive method about 10% more accurate and rapid than the equal additions method. It is easy to account for the advantage of the equal additions method over the decomposition method on account of its requiring less concentration; but it is not so easy to account for the 10% difference between the additive and the equal additions method. Here it would seem that the amount of concentration is about the same, for in each case the child sees part or all of the minuend, and in each method the subtrahend digit is increased by 1 under the same conditions.[2]

To summarize, all the available evidence seems to be definitely in favor of the Austrian method as the most efficient and the most easily taught procedure in subtraction; the equal additions method comes next in order of merit and should be the choice of those who, for some reason, are opposed to the Austrian method. The decomposition method, in view of the evidence now at hand, makes a poor showing, being inferior in both speed and accuracy to either of the other methods.

[1] Referring to the present writer's study of 1924, discussed in this volume on page 31.

[2] Stone, J. C. *How We Subtract*, pp. 68, 69, 74. Benj. H. Sanborn & Co., Chicago, 1926.

BIBLIOGRAPHY

ADAMS, DANIEL. *The Scholar's Arithmetic.* Revised Edition. John Prentiss, Keene, N. H., 1815.

ALMACK, JOHN C. *Research and Thesis Writing.* Houghton Mifflin Co., Boston, 1930.

BAKER, HUMPHREY. *The Wellspring of Sciences.* Thomas Purfoote, London, 1568 (1591 ed.).

BALLARD, P. B. "Norms of Performance in the Fundamental Processes of Arithmetic." *Journal of Experimental Pedagogy,* Vol. II, pp. 396–405, December, 1914; also Vol. III, pp. 9–20, March, 1915.

BEATTY, W. W. "The Additive versus the Borrowing Method of Subtraction." *Elementary School Journal,* Vol. XXI, pp. 198–200, November, 1920.

BLYTHE, JAS. E. *Complete Arithmetic.* Indiana School Book Co., Indianapolis, 1889.

BOND, ELIAS A. *Arithmetic for Teacher-Training Institutions.* Bureau of Publications, Teachers College, Columbia University, New York, 1934.

BOWERS, H. "Methods of Subtracting." *The School,* Vol. XXI, pp. 7–12, 105–108, September, October, 1932.

BREED, F. S., OVERMAN, J. R., AND WOODY, CLIFFORD. *Child-Life Arithmetics.* Lyons and Carnahan, Chicago, 1936.

BREUCKNER, L. J., ANDERSON, C. J., BANTING, G. O., AND MERTON, ELDA L. *The New Curriculum Arithmetic.* John C. Winston Co., Philadelphia, 1935.

BROWN, J. C., MIRICK, HELEN C., GUY, J. F., AND ELDREDGE, A. C. *Champion Arithmetics.* Row, Peterson and Co., Evanston, Ill., 1933.

BUCKINGHAM, B. R. AND OSBURN, W. J. *Searchlight Arithmetics. Four-Book Series.* Ginn and Co., Boston, 1927.

BUCKINGHAM, B. R. "The Additive versus the Take-Away Method of Teaching the Subtraction Facts." *Educational Research Bulletin,* No. 6, pp. 265–269, Ohio State University, 1927.

CAJORI, FLORIAN. *A History of Mathematics.* Macmillan Co., New York, 1931.

CHÂTELET, ALBERT. *Arithmétique, Cours Moyen.* Bourellier et Cie., Paris, 1934.

CIRODDE, P. L. *Leçons d'Arithmétique.* L. Hachette et Cie., Paris, 1858.

CLARK, JOHN R., OTIS, A. S., HATTON, CAROLINE. *First Steps in Teaching Number.* World Book Co., Yonkers, N. Y., 1929.

COCKER, EDWARD. *Decimal Arithmetick.* John Hawkins, London, 1684 (1703 ed.).

COLBURN, WARREN. *First Lessons in Intellectual Arithmetic.* Houghton Mifflin Co., Boston, 1821.

COLBURN, WARREN. *Intellectual Arithmetic.* W. J. Reynolds and Co., Boston, 1821 (1847 ed.).

COLEBROOKE, HENRY THOMAS, ESQ. *Algebra with Arithmetic and Mensuration, from the Sanscrit of Brahmequpta and Bhascara.* John Murray, London, 1817.

DABOLL, NATHAN. *School Master's Assistant.* Mack, Andrus and Woodruff, Ithaca, N. Y., 1797 (1837 ed.).

DANIEL, MARGARET D. "Comparative Merits of Four Methods of Teaching Subtraction." Unpublished Master's Thesis, University of California, August, 1931.

DAVIES, CHARLES. *University Arithmetic.* American Book Co., New York, 1864.

DE MORGAN, AUGUSTUS. *Arithmetical Books from the Invention of Printing to the Present Time.* Taylor and Walton, London, 1847.

DE MORGAN, AUGUSTUS. *Elements of Arithmetic.* J. Taylor, London, 1835.

DE MORGAN, AUGUSTUS. *Elements of Arithmetic.* Walton and Moberly, London, 1854 ed.

DILWORTH, THOMAS. *The Schoolmaster's Assistant.* Jos. Crukshank, Philadelphia, 1781.

DURELL, FLETCHER AND GILLET, H. O. *The New Day Arithmetics.* Charles E. Merrill Co., New York, 1930.

EMERSON, FREDERICK. *North American Arithmetic.* Russell, Odiorne, and Metcalf, Boston, 1834.

FELTER, S. A. *Primary Arithmetic.* Charles Scribner and Co., New York, 1863.

FISHER, GEORGE. *The American Instructor.* John Bioren, Philadelphia, 1775.

FRISIUS, GEMMA. *Arithmeticae Practicae, Methodus Facilis.* Antwerp, 1540 (1581 ed.).

GARRETT, H. E. *Statistics in Psychology and Education.* Longmans, Green and Co., New York, 1926.

HALSTED, G. B. *Foundations and Technique of Arithmetic.* The Open Court Publishing Co., Chicago, 1912.

HARVEY, L. D. *Practical Arithmetic.* American Book Co., New York, 1908.

JACKSON, L. L. *The Educational Significance of Sixteenth Century Arithmetic.* Bureau of Publications, Teachers College, Columbia University, New York, 1906.

JOHNSON, J. T. "The Merits of Different Methods of Subtraction." *Journal of Educational Research,* Vol. X, No. 4, pp. 279-290, November, 1924.

JOHNSON, J. T. "The Efficiency of Decomposition Compared with That of Equal Additions as a Technique in Subtraction of Whole Numbers." *Mathematics Teacher,* Vol. XXIV, pp. 5–13, January, 1931.

JUDD, C. H. *Psychological Analysis of the Fundamentals of Arithmetic.* University of Chicago Press, Chicago, 1927.

KARPINSKI, L. C. *The History of Arithmetic.* Rand McNally and Co., Chicago, 1925.

KARPINSKI, L. C. "Two Twelfth Century Algorisms." *Isis,* Vol. III, pp. 396–413, 1921.

KNIGHT, F. B., RUCH, G. M., STUDEBAKER, J. W., AND FINDLEY, W. C. *Study Arithmetics.* Scott Foresman & Co., Chicago.

LANGE, GERSON. *Die Praxis des Rechners.* L. Golde, Frankfort, 1909. A translation from the Hebrew of Levi ben Gershom's Sefer Maassei Chosecheb of 1321.

LEMOYNE, ALCIDE AND AYMARD, AUBIN. *La Théorie Arithmétique.* H. Paulin et Cie., Paris, 1910.

LINDQUIST, E. F. "The Significance of a Difference Between 'Matched' Groups." *Journal of Educational Psychology,* Vol. 22, pp. 197–204, March, 1931.

McCALL, WILLIAM A. *How to Measure in Education.* Macmillan Co., New York, 1923.

McCLELLAN AND AMES. *Public School Arithmetic.* Macmillan Co., New York, 1902.

McCLELLAND, W. W. "An Experimental Study of the Different Methods of Subtraction." *Journal of Experimental Pedagogy,* Vol. IV, pp. 293–299, December, 1918.

McCURDY, D. *New American Order of Arithmetic.* Armstrong & Berry, Baltimore, 1850.

MEAD, C. D. AND SEARS, I. "Additive Subtraction and Multiplicative Division Tested." *Journal of Educational Psychology,* Vol. VII, pp. 261-270, May, 1916.

MILL, JOHN STUART. *System of Logic,* Vol. I, Book III. Harper and Brothers, New York, 1850 ed.

MILNE, WM. J. *Standard Arithmetic.* American Book Co., New York, 1895.

MORTON, R. L. *The Teaching of Arithmetic in the Primary Grades.* Silver, Burdett and Co., New York, 1927.

MORTON, R. L. *Teaching of Arithmetic in the Intermediate Grades.* Silver, Burdett and Co., New York, 1927.

MORTON, R. L. *Teaching Arithmetic in the Elementary School.* Silver, Burdett and Co., New York, 1937.

MURRAY, JAMES A. H. *A New English Dictionary,* Vol. I, A and B. Oxford University Press, Oxford, England, 1888.

NATIONAL COUNCIL OF TEACHERS OF MATHEMATICS. *The Tenth Year Book, The Teaching of Arithmetic.* Bureau of Publications, Teachers College, Columbia University, New York, 1935.

OSBURN, W. J. "How Shall We Subtract?" *Journal of Educational Research,* Vol. XVI, pp. 237–246, November, 1927.

PIKE, NICHOLAS. *A New and Complete System of Arithmetic.* J. Mycall, Newburyport, Mass., 1788.

RAY, JOSEPH. *Higher Arithmetic.* American Book Co., New York, 1880.

RECORDE, ROBERT. *The Whettstone of Witte.* John Kyngstone, London, 1542 (1557 ed).

RIESE, ADAM. *Rechenung nach der lenge auff den Linihen und Feder.* Jacobum Berwalt, Leipzig, 1522 (1550 ed.).

ROANTREE, W. F. "The Question of Method in Subtraction." *Mathematics Teacher,* Vol. XVII, pp. 110–116, February, 1924.

ROBINSON, ARTHUR E. *The Professional Education of Elementary Teachers in the Field of Arithmetic.* Bureau of Publications, Teachers College, Columbia University, New York, 1936.

ROBINSON, D. S. *The Principles of Reasoning.* D. Appleton and Co., New York, 1924.

ROBINSON, H. N. *Practical Arithmetic.* Ivison, Blakeman, Taylor and Co., New York, 1881.

SANFORD, VERA. *A Short History of Mathematics.* Houghton Mifflin Co., Boston, 1930.

SMITH, DAVID EUGENE. *History of Mathematics,* Vol. II. Ginn and Co., Boston, 1925.

SMITH, DAVID EUGENE. *Rara Arithmetica,* 2 vols. Ginn and Co., Boston, 1908.

SMITH, DAVID EUGENE. *Source Book in Mathematics.* McGraw-Hill Book Co., Inc., New York, 1929.

SMITH, D. E., LUSE, EVA MAY, AND MORSS, E. L. *Problem and Practice Arithmetics.* Ginn and Co., Boston, 1929.

SPEER, W. W. *Elementary Arithmetic.* Ginn and Co., Boston, 1902.

SPICKERNELL, G. E. *An Explanatory Arithmetic.* Griffin and Co., London, 1883.

STAUBACH, CHARLES N. "An Anglo-Norman Algorism of the Fourteenth Century." With Introduction by L. C. Karpinski. *Isis,* Vol. XXIII, pp. 121–152, 1935.

STEELE, ROBERT. *The Earliest Arithmetics in English.* H. Milford, London, 1922.

STIFEL, MICHAEL. *Die Coss Christoffs Rudolffs.* W. Jansen, Amsterdam, 1615.

STONE, J. C. *The Stone Arithmetic, Third Year.* Benj. H. Sanborn & Co., Chicago, 1929.

STONE, J. C. *How We Subtract.* Benj. H. Sanborn & Co., Chicago, 1926.

STONE, J. C. AND MILLS, C. N. *Unit Mastery Arithmetic. Three-Book Series.* Benj. H. Sanborn & Co., Chicago, 1932.

STRAYER, G. D. AND UPTON, C. B. *Strayer-Upton Arithmetics. Three-Book Series.* American Book Co., New York, 1928.

STUYVAERT, M. *Les Nombres Positifs.* Van Goethem et Cie., Gand, Belgium, 1912.

TACQUET, ANDRÉ. *Arithmeticae, theoria et praxis.* I. Meursium, Antwerp, 1665.

TANNERY, JULES. *Leçons d'Arithmétique et Pratique.* Colin et Cie., Paris, 1900.

TAYLOR, J. S. "Subtraction by the Additive Process." *Elementary School Journal,* Vol. XX, pp. 203–207, November, 1919.

THOMPSON, J. B. *Arithmetic Series.* Newman and Ivison, New York, 1847.

THORNDIKE, EDWARD LEE. *The New Methods in Arithmetic.* Rand McNally and Co., Chicago, 1921.

THORNDIKE, EDWARD LEE. *The Psychology of Arithmetic.* Macmillan Co., New York, 1923.

THORNDIKE, E. L. *The Thorndike Arithmetics.* Rand McNally and Co., Chicago, 1917 (1924 ed.).

Twenty-ninth Yearbook of the National Society for the Study of Education. Public School Publishing Co., Bloomington, Ill., 1930.

WALKER, H. AND DUROST, W. *Statistical Tables, Their Structure and Use.* Bureau of Publications, Teachers College, Columbia University, New York, 1936.

WALSH, JOHN H. *Practical Methods in Arithmetic.* D. C. Heath and Co.. Boston, 1911.

WATERS, E. G. R. "A Fifteenth Century French Algorism from Liége." *Isis,* Vol. XII, pp. 194–236, 1929.

WATERS, E. G. R. "A Thirteenth Century Algorism in French Verse." With Introduction by L. C. Karpinski. *Isis,* Vol. XI, pp. 45–84, 1928.

WENTWORTH, G. A. AND SMITH, D. E. *Complete Arithmetic.* Ginn and Co., Boston, 1909.

WENTWORTH, G. A. *Elementary Arithmetic.* Ginn and Co., Boston, 1895.

WHEAT, G. H. *The Psychology and Teaching of Arithmetic.* D. C. Heath and Co., Boston, 1937.

WILKS, S. S. "The Standard Error of the Means of Matched Samples." *Journal of Educational Psychology,* Vol. 22, pp. 205–208, March, 1931.

WILSON, G. M. "For 100% Subtraction, What Method? A New Approach." *Journal of Educational Research,* Vol. XXVII, pp. 503–508, March, 1934.

WINCH, W. H. "Equal Additions versus Decomposition in Teaching Subtraction." *Journal of Experimental Pedagogy,* Vol. V, pp. 207–220, 261–270, June, 1920.

WINGATE, EDMUND. *Arithmetick, Containing a Plain and Familiar Method for Attaining the Knowledge and Practice of Common Arithmetick.* Edited by James Dodson, N. P. London, 1629 (1760 ed.).

YULE, G. U. *An Introduction to the Theory of Statistics.* C. Griffin and Co., Ltd., London, 1929.